PLANT SCIENCE RESEARCH

PINUS

GROWTH, DISTRIBUTION AND USES

PLANT SCIENCE RESEARCH AND PRACTICES

Additional books and e-books in this series can be found on Nova's website under the Series tab.

PLANT SCIENCE RESEARCH AND PRACTICES

PINUS

GROWTH, DISTRIBUTION AND USES

SYLVESTER STEPHENS
EDITOR

nova
science publishers
New York

NOTICE TO THE READER

Library of Congress Cataloging-in-Publication Data

ISBN: 978-1-53616-429-9

Published by Nova Science Publishers, Inc. † New York

CONTENTS

PREFACE

In *Pinus: Growth, Distribution and Uses*, analyses of the current state of Scots pine (*Pinus sylvestris* L.) and the distribution of pine forests was carried out in three zones of East European plain: a taiga, a temperate broadleaf and mixed forest, and a temperate steppe.

A study is included which focuses on somatic embryogenesis in Yakutanegoyou, *Pinus armandii* Franch. var. amamiana (Koidz.) Hatusima, an endemic and endangered species in Japan, which was initiated from megagametophytes containing zygotic embryos on a medium supplemented with 2,4-dichlorophenoxyacetic acid and 6-benzylaminopurine.

Lastly, the authors gather and present information about pinus cultivation, as well the potential use of pinus bark as an adsorbent to remove $Cd2+$, $Pb2+$ and $Cr3+$ from contaminated water.

Chapter 1 – Analyses of the current state of the Scots pine (*Pinus sylvestris* L.) and the distribution of pine forests was carried out in the three zones of the East European (Russian) Plain: a taiga, a temperate broadleaf and mixed forest, and a temperate steppe. This is the unique natural landscape, with a relatively homogeneous ecotope and biocenosis. Nearly 70% of forest lands have the optimal geomorphological, hydrothermal, and soil conditions. In this Chapter, the experimental data of the changes in the forest cover, its location, and the proportion of pine forests along the three latitudinal transects (from East to West) are discussed. The data on deforestation and reduction of the area of pine forests during the past 300 years are given. It is shown that the natural forests within the temperate broadleaf and mixed forest, and temperate steppe zones were almost totally destroyed. Currently, the most acute problem is their artificial reproduction, optimization of the age structure of stands, and the increase of productivity and sustainability to the abiotic and biotic impacts. Distribution of pine forests by the latitudinal vector of the East European Plain was carried out by means of two indicators of population instability – interspecies competitiveness and quality of pine-tree subgrowth at a natural reforestation. It has been established that the vital state of pine forests changes twice in the direction from north to south of range. The temperate broadleaf and mixed forest zone is included in the territory of highest instability of the species. On this region the processes of natural reforestation, as a rule, proceed with a change of species due to the low quality of pine-tree subgrowth. Because of the decrease in interspecies competitiveness, the Scots pine has been gradually replaced from forest cover mainly by birch. This phenomenon is currently observed along all latitudinal transects and is increasing from east to west. We came to the conclusion, that there has been the break of range in half. It passes through the center of the temperate broadleaf and mixed forest zone of the East European Plain and divides the species range into northern (taiga) and southern (temperate steppe) parts. The different reasons for destabilization are analyzed and discussed. It is shown that anthropogenic load promotes the expansion and deepening of discontinuity region, but is not its cause. It is assumed that destabilization of the pine gene pool is due to the peculiarities of the evolution of forest cover as a whole and the pine evolution, as the previous edificatory species, one of the links of which is the weakening of connections in the "parents-progeny" system.

Chapter 2 – Somatic embryogenesis in Yakutanegoyou, *Pinus armandii* Franch. var. *amamiana* (Koidz.) Hatusima, an endemic and endangered species in Japan, was initiated from megagametophytes containing zygotic embryos on a medium supplemented with 2,4-dichlorophenoxyacetic acid and 6-benzylaminopurine. Embryogenic cultures were maintained and proliferated by subcultures at 2- to 3-week intervals on the same fresh medium. The maturation of somatic embryos occurred on media containing maltose, polyethylene glycol, abscisic acid, and activated charcoal. High frequencies of germination were obtained after the post-maturation treatment of somatic embryos under conditions of high relative humidity, and after the maturation of somatic embryo on medium with a high concentration of gellan gum.

Chapter 3 – The intensification of human activities generates an increasing demand for products of forest origin. The *Pinus* sp. genus, for example, has been extensively explored and consequently, it has generated a large amount of wastes (barks and sawdust from wood extraction). If they are not well managed, the barks generated by pinus can be harmful to the environment, since they store large amounts of toxic metals. In this context, the destination of the barks to obtain adsorbents for remediate toxic metals from contaminated water can be an alternative. Therefore, the present study aimed to gather information about the pinus cultivation, as well the potential use of the pinus bark as adsorbent to remove Cd^{2+}, Pb^{2+} and Cr^{3+} from contaminated water. It was done a search about the pinus cultivation in the world, focusing for studies that aimed to produce adsorbents from pinus bark and the results obtained in the remediation of contaminated water. Based in the data collection in this work, it can be conclude that the modified pinus barks become an excellent alternative in the removal of toxic metals from water (Cd^{2+}, Pb^{2+} and Cr^{3+}). In addition, the use of pinus bark as an adsorbent material represents a sustainable practice (appropriate destination of the pinus bark and the decontamination of the water), which complements the final productive steps of this species. Moreover, the adsorption process is deeply influenced by chemical compounds that substantially change the characteristics of the material and increase the adsorption.

In: Pinus
Editor: Sylvester Stephens

ISBN: 978-1-53616-429-9
© 2019 Nova Science Publishers, Inc.

Chapter 1

PINE FORESTS OF EAST EUROPEAN PLAIN: DISTRIBUTION TRENDS, FUNCTIONS AND DEVELOPMENT PROBLEMS

Nina F. Kuznetsova, Michael A. Semenov
and Marina Yu. Sautkina[*]
Federal Forestry Agency of the Russian Federation,
All-Russian Research Institute of Forest Genetics,
Breeding and Biotechnology, Voronech, Russia

ABSTRACT

Analyses of the current state of the Scots pine (*Pinus sylvestris* L.) and the distribution of pine forests was carried out in the three zones of the East European (Russian) Plain: a taiga, a temperate broadleaf and mixed forest, and a temperate steppe. This is the unique natural landscape, with a relatively homogeneous ecotope and biocenosis. Nearly 70% of forest lands have the optimal geomorphological, hydrothermal, and soil conditions. In this Chapter, the experimental data of the changes in the forest cover, its location, and the proportion of pine forests along

[*] Corresponding Author's E-mail: nfsenyuk@mail.ru.

the three latitudinal transects (from East to West) are discussed. The data on deforestation and reduction of the area of pine forests during the past 300 years are given. It is shown that the natural forests within the temperate broadleaf and mixed forest, and temperate steppe zones were almost totally destroyed. Currently, the most acute problem is their artificial reproduction, optimization of the age structure of stands, and the increase of productivity and sustainability to the abiotic and biotic impacts. Distribution of pine forests by the latitudinal vector of the East European Plain was carried out by means of two indicators of population instability – interspecies competitiveness and quality of pine-tree subgrowth at a natural reforestation. It has been established that the vital state of pine forests changes twice in the direction from north to south of range. The temperate broadleaf and mixed forest zone is included in the territory of highest instability of the species. On this region the processes of natural reforestation, as a rule, proceed with a change of species due to the low quality of pine-tree subgrowth. Because of the decrease in interspecies competitiveness, the Scots pine has been gradually replaced from forest cover mainly by birch. This phenomenon is currently observed along all latitudinal transects and is increasing from east to west. We came to the conclusion, that there has been the break of range in half. It passes through the center of the temperate broadleaf and mixed forest zone of the East European Plain and divides the species range into northern (taiga) and southern (temperate steppe) parts. The different reasons for destabilization are analyzed and discussed. It is shown that anthropogenic load promotes the expansion and deepening of discontinuity region, but is not its cause. It is assumed that destabilization of the pine gene pool is due to the peculiarities of the evolution of forest cover as a whole and the pine evolution, as the previous edificatory species, one of the links of which is the weakening of connections in the "parents-progeny" system.

Keywords: optimal range zone, southern taiga, temperate broadleaf and mixed forest, temperate steppe, population instability, interspecies competitive ability, pine regeneration

1. Characteristic of East European Plain, as the Research Object

The East European Plain is the 2nd largest plain on the Planet. Its geographical position is in the eastern part of Europe (Figure 1). The average height is about 170 m above sea level. The East European Platform forms its basis, the geological structures of which determine the main features of its relief. It is one of the most stable areas of the earth's crust with a Precambrian crystalline basement. Quaternary glaciation for its northern half had a large relief-forming role. Here the Plain territory is characterized by hummock-and-hollow topography with large number of lakes. This type of terrain in more southern areas is found only in places between the flat interfluves.

The territory of the Plain is a unique landscape despite different climatic conditions and heterogeneity of the surface topography. Its relief consists mainly of elevated and low-lying areas, as well as plains and terraces in river valleys. The subdued terrain provides mainly western disturbance of air masses. The Atlantic air in the summer brings rain and cool weather, and in the winter brings rainfalls with warming weather. The continentality of the climate increases from northwest to southeast. The change of the main components of the Plain (neutral moisture balance and border of loess-like soils) and its topography occurs closer to the temperate steppe zone (Nesterov and Fedotov, 2005). The southern part is characterized by more rugged terrain and consists of slight plateau-like elevations and low-lying areas, wide valleys, and, as a rule, is separated by beams and ravines (Neustruev, 1930).

At present, the maps of forest cover development in Eastern Europe for more than 10,000 years have been created on the basis of biological and archaeological data (Kozharinov, 1995; Bondarev, 1998). It is shown that a species composition of plain forests throughout the post-glacial-Holocene period has changed with a frequency of 3.5 thousand years, 2.5-1.0 thousand years, and 500 years (Smirnova and Bobrovsky, 2000). The last formation of zonation, species composition, and boundaries of ranges on the territory of European Russia was formed one or two thousand years ago. More than half of the dry lands in the central part are characterized by favorable forest conditions. The geography of the location of ecological trouble zones (swamps, mountains, and northern and southern borders of the range) has not changed during many centuries. The general specificity of plain forests, formed after glaciation, is their evolutionary youth, poor species composition (21 species, more than 80% of forest lands are occupied by 11 species), widespread edificatory species with a strategy of rapid growth and reproduction, and resistance to ecocide effects (Isaev et al., 1995; Romanovsky, 2002; Zhigunov, 2008).

The East European Plain is characterized by a relatively homogeneous biocenosis and ecotope (Rysin, 1995). The relief and climate of the Plain determine a clear-cut change of natural zones from north to south. Forests cover more than half of its territory. They grow in the taiga, temperate broadleaf and mixed forest, temperate steppe, and steppe zones. The taiga is divided into the northern, middle, and southern subzones (Milkov, 1977). Taiga (northern, middle) and steppe are the regions of the northern and southern pessimum of forest cover. Southern taiga, temperate broadleaf and mixed forest, and temperate steppe zones occupy the central part of Plain, where the complexes of the best physiographic and biocenotic conditions, which is necessary for the formation of boreal, deciduous, and mixed forests, are concentrated. Here, the largest forest woodlands (simple, complex in species composition) grow. Their populations are on average greater, and the completeness of forest stand is the higher. Zones of species optimum of edificators such as spruce, pine, birch, oak, linden, etc. pass through this territory. The forests perform many functions: resource and raw materials, environment-forming, soil-protective, water-regulating, and health-improving.

The analysis of tree cover showed that the plain forests with average productivity and average species composition(6-11 species) are the most stable and productive (Romanovsky, 2002).Stability in the "rich" and "poor" variants, as a rule, is lower. Forests of the taiga and temperate steppe zones are an exception. There are one-, two-, and three- species forest massifs characterized by stability, competitiveness, and productivity (Mamaev and Makhnev, 1986). Changes of species composition (digressive changes of 1, 2 orders) after cutting down or fire often occur in them, the final result of which is the achievement of the equilibrium state of the initial forest. The duration and direction of the succession rows directly depend on the degree of deviation of the derived forest from the initial ones (Vysotsky, 1962).

At the end of the twentieth century, the principle of the ecological-genetic commonality of populations within ecotopes was substantiated (Schwartz, 1980; Glotov, 1983). According to this principle, different species have a set of common norm reactions, due to which they react the same way (down to the details) to the similar environmental conditions. The species exists until the gene pool of species norm is lost and the habitat is preserved, where it can be realized in the population of the phenotypic and genotypic norm. The complex of favorable climatic and edaphic conditions is concentrated in the center of the ranges. Species in zones of ecological disadvantage can wait out unfavorable periods, but not evolve. Most forests in the center of the East European Plain are in equilibrium, which ensures their stability as structural units of nature (Romanovsky, 2002). For one or two millennia, more than one generation of forests has changed on its territory. Multiple selections in the same environment had formed the forests, which are adapted to the climate and terrain according to the principle of interaction "species-environment"– "genotype-environment." The life of the forest and the life of animals, plants and microorganisms under its cover are joined into a single complex. The constituent parts were in a state of dynamic equilibrium for a long time.

Any type of biotic and abiotic stress throws off balance in the genetic system of trees and forest ecosystems (Levitt, 1980; Romanovsky, 2002; Harfouche et al., 2014; etc.). Instability is a particular state which negatively affects their growth and productivity. Fire has historically been the mainfactor, which has led to direct forest destruction. New forests arose at the sites of the burning, being formed after fires. This contributed to the renewal of forest cover and the mixing of genetic material throughout the range.

The mismatch between the forest gene pool and environment manifests as instability at the population and individual levels (Romanovsky, 2002). The result are a loss of such properties as integrity and continuity, a resumption of forests with changing rocks, a reduction in the competitiveness of trees and species, a destabilization of the generative sphere, violations in the system of "parents-progeny", and an inability of the species in places of traditional growth to produce high-quality progeny and a low its survival (Isakov et al., 1989; Romanovsky, 2002; Rysin, 2008; Kuznetsova, 2010; 2012; 2019).

From the beginning of the 2nd millennium, a new wave of human settlement began across the East European Plain (Pushkova, 1968; Abaturov, 1995).Human impact on the forests is manifested in the genetic, ecological, and geographical aspects. During the XVIII-XX centuries, the area of natural forests on the territory of the East European Plain was constantly decreasing, and their structure was destroyed. In this period, almost a third of the plain forests were lost (Tsvetkov, 1957). If in 1696 the forest cover was 52.7%, then by 1914 it had decreased to 35.2%. The process of deforestation from north to south was uneven. The process of deforestation from north to south was uneven. In relation to the optimal level, the forest cover of the taiga zone decreased by only 20-30%, and temperate broadleaf and mixed forest by 40-60%.In the temperate steppe zone only 1/3 of the initial forests remained. The first half of the XX century was characterized by minimal forest cover (23.5%). Riparian or flooded forests suffered the most, their area decreasing by 3-5 times. The natural forest cover was gradually replaced by anthropogenic (Kuznetsova and Sautkina, 2019).

To the middle of the XX century, there were almost no forests not affected by economic activity in the temperate broadleaf and mixed forest and the temperate steppe zones. The remaining forests were highly fragmented, characterized by disturbed flora and fauna and species and age composition. Only in the taiga zone remained large massifs of natural coniferous forests (Pisarenko, Strakhov, 2004, 2014). In the 1950s – 1990s, large-scale reforestation began in the country (Pisarenko, 1977; Redko, Treschevsky, 1986). Over the years, a forest culture was created on an area of 42 million hectares. More than 70% of them were located in the central and southern parts of the East European Plain. When creating plantations, preference was given to pine and spruce in the temperate broadleaf and mixed forest zone and pine and oak in the temperate steppe ones.

Many beams and ravines were forested. Systems of forest shelter belts, large forest belts along rivers and on the watersheds between them, was created.

Pine forests belong to a zonal type of vegetation. They cross the boundaries of latitudinal natural zones from the taiga to steppe region of the *Pinus sylvestris* range. This is due to the fact that the pine tree can tolerate both the harsh climate of the north and the hot climate of the steppe.

Pine is an edificatory species of both the plain and mountain light-needle coniferous forests of European Russia (Bulygin, 2001). This tree species has high-quality wood. In some regions of Russia, resin tapping is still carried out. It is important for chemical industry development. The ecological function of this species is also significant. Pine releases special substances, phytoncides, which purify the air from various harmful organisms. Conifers accumulate solar energy better than other species of woody and grassy plants. Pine is widely used in the field of protective forestation. It is the main species for the creation of forest plantations. Pine forests improve the gas composition of the atmosphere by means of carbon accumulating in the phytomass of trees (Filipchuk and Moiseev, 2001). The role of pine plantations in the global carbon cycle is given in Table 1-2 by the case of steppe and forest-steppe regions of East European Plain (Development of scenarios ..., 2017).

Table 1. Carbon stocks of forest ecosystems in the region of forest-steppe and steppe of East European Plain (forest-covered lands)

Dominant species	Area, thousand hectares	Stock, million м^3	Carbon, million t of C					
			tree stand	vegetable cover	dead wood	litter	soils (0-30 см)	subtotal
Scots pine	893.8	158.93	46.9762	0.4785	15.4175	7.4067	62.9736	133.253
Subtotal (conifers)	912.8	164.6	47.5792	0.48423	15.5435	7.5653	64.4039	135.576
Subtotal (forest-covered lands)	4015.9	588.64	195,864	2.10948	47.436	46.4927	243.874	535.776

Table 2.Carbon sequestration by forest ecosystems in the forest-steppe and steppe of East European Plain (forest-covered lands)

Dominant species	Area, thousand hectares	Stock, million м³	Carbon, million t of C				
			tree stand	dead wood	litter	soils (0-30 см)	subtotal
Scots pine	893.8	162.98	1083.1	350.7	75.4	280.9	1790.1
Subtotal (conifers)	912.8	164.6	1110.1	355.2	79.4	311.5	1856.1
Subtotal (forest-covered lands)	4015.9	588.64	4278.5	1033.6	180.6	610.5	6103.2

In the XXI century, forest eliminating occurs by means of destruction of the gene pool and/or its habitat. The second path, the conditions of global climate change and environmental crisis, is developing more rapidly. Over the past 25 years, the area of fires has doubled and there is an increase in the number of large fires. The upper soil horizons are compacted, and many physical and chemical properties of the soil change, without which the normal development of roots and soil microflora is impossible. The chemicalization of agriculture and the massive use of chemicals in everyday life results in these substances, after the melting of snow, falling into rivers, the soil, and groundwater. Note that the consequences of the environment destruction are more dangerous than the elimination of part of the gene pool. The volume of the gene pool after the normalization of the environment can be increased. The technologically transformed environment is more difficult to recover and this takes more time. Its continued destruction threatens the possibility of having quality forests in the future.

The general conclusion is that most of the forests of the East European Plain have been created in the result of human activities and are the derivative by origin. A positive factor is that the Plain ecotope, with which they are connected by many trophic, edaphic, and adaptive connections, has not yet lost its potential. Since the habitat is preserved, most of the forests in the center of the species ranges are characterized by relatively high bonitet, and greater longevity and resistance to a set of major regional stressors.

As is known, the ability of natural regeneration is one of the essential components of forest life. When comparing the current state of pine forests, we used such a trait of population instability as degree of violations in the system of "parents-progeny" (Isakov et al., 1989; Kuznetsova and Sautkina, 2019). It is expressed as a loss of yield and a reduction in the quality and quantity of seed progeny in places of traditional growth of *Pinus sylvestris*. An analysis of the current state of pine forests in order to identify the geography and causes of the nascent gap in the range of this species will be carried out in the three zones of the East European Plain (Figure 1): from east to west – a southern taiga, a temperate broadleaf and mixed forest and a temperate steppe zones.

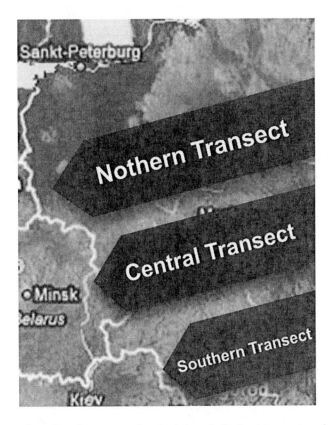

Figure 1. Location of northern, central and southern latitudinal transect on the territory of East European Plain.

It should be ensured that a Scots pine is one of the most light-loving species. Therefore, it is very important to note that its attitude towards the light changes with age. The pine is most shade-tolerant in the first years of life. At this time, the composition and fertility of the soil have a significant impact on its shade tolerance, since with the best supply of water and nutrients the needles absorb most of the light falling on them (Bulygin, 2001). The seed progeny of pine (or pine-tree subgrow thunder the forest canopy) with the same illumination is more depressed, then the soil becomes the more and drier and poorer.

2. NORTHERN LATITUDINAL TRANSECT: SOUTHERN TAIGA SUBZONE

Scots pine (*Pinus sylvestris* L.) has a continuous range. This species, according to Bobrov (1978), appears on the Russia territory from the Miocene and it spreads from west to east. Its outline has changed several times due to climate change. The range of pine 10-12 thousand years ago was half as much as the modern pine (Neustadt, 1964; Pokrovskaya, 1966). The last ice age had a great influence on its development. The age of the current range is estimated at 4-6 thousand years (Pravdin, 1964). The northern extent of its boundary is 67 degrees north latitude and almost coincides with the boundary of forest cover. The southern ones are located at 37 degrees north latitude.

A peculiarity of northern taiga is a small amount of solar radiation. Forests on its territory occupy significant areas (Mamaev, 1973).This is the northern boundary of the Scots pine range. Pine forests here are usually waterlogged, are of low completeness (0.2-0.4), and are of a low bonitet class (IV-V). The low temperature regime inhibits the development of trees, slowing the growth of roots and making it difficult to assimilate nutrients (Rysin, Savelieva, 2008). They are characterized by a large number of dwarf and semi-dwarf trees (the forms with a curved trunk, one-sided crown, and high sterility). Trees are polymorphic in height, and have anatomical and morphological traits of needles and cones (Abaturova and Khromova, 1984).

The middle taiga is also the territory of Scots pine pessimum. The vegetation period is longer (105-120 days), the summer is warmer, and the winter is long-drawn-out and frosty. The forests are characterized by higher closeness and they are less swamped. If in the northern taiga to mature age, they accumulate between 50 and180 tons ha-1 of organic matter, and in the middle taiga accumulate between 126 and 269 tons ha-1 of organic matter (Bobkova, 1987).Pine forests in this subzone also do not achieve optimal vegetative and seed productivity (bonitet class III - IV).. They are characterized by such adaptations as a short period of growth and the passage of vital processes at lower temperatures.

Pine in the northern and middle taiga dominate over other edificatory species (Rysin and Savelieva, 2008). Its part of the total area of forests is 43%, with spruce and birch contributing 31% and 26% respectively. This ratio, as the most optimal proportions between tree species, is used as a control in our analysis. The area of pine forests has steadily increased from 2010 onwards at a rate of 15.9 thousand hectares per year. Reforestation occurs mainly through natural regeneration (72.3%). The pine-tree subgrowth under a pine canopy is developed in the form of aggregations or is distributed diffusely, is rather abundant, and corresponds to the category of "healthy" (Kucherov and Zverev, 2012). This indicates that the connection in the "parents-progeny" system is not broken despite the presence of anthropogenic pressure. Therefore, the probability of pine domination after disintegration of pine forests is high.

The object of our research is the pine forests of the southern subzone of taiga, which is characterized by a warmer climate. The average annual temperature is between 2.8 and 3.8°C. The duration of vegetative season is between 166 and170 days. The average annual rainfalls is between 560 and 640 mm. These climatic conditions are quite favorable for the development of coniferous plants. The southern taiga is characterized by a higher species diversity. The average productivity of forests is between II and III class of quality (Mamaev, 1973). It is territory of species optimum for the Scots pine range.

In the center of the species range, the gene pool of *Pinus sylvestris*, as the whole formation, has a number of common properties, such as uniformity and continuality (Kuznetsova, 2019). Here the zonal traits are poorly pronounced and the intraspecific differentiation is more due to the edaphic, climatic, and ecological conditions of region. For this reason, the inter population boundaries of pine forests correspond with the boundaries of physical-geographic areas and the boundaries of groups of populations coincide with the basic configurations of relief (Vidyakin, 2007). Scots pine in temperate broadleaf and mixed forest and temperate steppe areas forms highly productive pine forests. Its trees reach a height of 35-45 m and are the trees of the first magnitude. Pine in the open spaces begins to fructify at the age of 7-10 years, (15-30 years in plantations), and its seed production continues until old age. The seed formation in Scots pine occurs during 2 years and 8 months and it is a genetically determined and environmentally dependent process. In the taiga zone, its pollination occurs during May and June, when the amount of effective (above 5°C) temperatures reaches approximately 190-200°C (Artemov, 1985). In the temperate steppe zone, the pollination begins earlier (II–III May decades) with a sum of temperatures between 270 and 320°C (Svintsova et al., 2014; Kuznetsova, 2009, 2019).Fertilization takes place at the following year. In the forest-steppe it is during the I - II decades of June, with sum of effective temperatures between510 and 550°C. The average life expectancy of trees is 300-400 years. The most significant increase in height occurs between the ages of 10 and 40 years (Bulygin, 2001).

The central populations of the now existing plain pine forests practically do not differ by the phenotypic, cytogenetic, and molecular genetic traits (Sannikov and Petrova, 2003). The study of cones and seeds by 24 signs showed small average (pair-group) phenotypic distances of Mahalanobis between groups of central populations of the East European Plain (Sannikov et al., 1997). Variability of pollen grains, micro- and megastrobiles, seeds, and anatomic elements of needles is also insignificant (Semeryakov, 1986; Sannikov and Petrova, 1997, 2003; Vidyakin, 2014). Differences in the timing of pollination in biogeocenoses do not exceed 1-2 days. Degree of phenological isolation is 10-15%

(Sannikov and Petrova, 2007). Level of mitosis pathologies is within 1-5% (Butorina et al., 2001; Pardaeva et al., 2013). The inter population variability by molecular genetic traits is in the range of 2–6% (Goncharenko and Silin, 1997; Sannikov et al., 2002; Sannikov and Petrova, 2003).

The northern transect from east to west passes through southern subzone of the taiga zone along the East European Plain (Figure 1). Its length is equal to 1100 km and includes three regions: the eastern, the central, and the western. A significant part of this territory is covered with pine-spruce, pine-birch, and birch forests. Here are the northern boundaries of optimal zones for the European spruce, Scots pine, and Silver birch, which account for 90% of forest land. Scots pine, as a stress-tolerant species, grows on different soils, in plains, and in the mountains, and is characterized by high environmental plasticity and occupies vast territories. However, a pine in the mountains, as a rather heat-loving species, never reaches the upper boundary of the forest. Pine-spruce forests from II to V bonitet are spread mainly on depleted soils (Kurnaev, 1973). Pine forests with spruce of I – II bonitet are formed on rich soils. Pure pine forests of IV - II bonitet are grown on sandy soils.

The starting point of our analysis is the Kostroma region, which is completely located in the southern taiga and has a length of 420 km (from east to west). The relief of its territory is a hilly plain, which was formed in the Quaternary period, during the Dnieper and Moscow glaciations (Dudin, 2000). Sod-podzolic (55% of territory) and marsh-podzolic (11%) soils are predominant. The climate is temperate continental. The average monthly temperature for June-August is 21-23°C.

Climatic and site characteristics are favorable for the development of productive coniferous (spruce, pine - II - III bonitet class) forests and highly productive birch forests. Until the 19th century, the share of conifers was 75% (Dudin, 2000, 2011). The forest expanses were huge and are still characterized by the largest stock of wood (the average stock per 1 ha is 169 m^3). Forest cover is still very high – (74.2%). Pine forests occupy mostly sandy and sandy lowland plains. Natural renewal is abundant (1.8-2.0 thousand/ha). It is successfully carried out in pine forests and in 50% of logging areas (Lviv et al., 1980). Pine undergrowth under the forest canopy is characterized by high quality and uniform placement (Bagaev, 2018).

The extension of agricultural land and the timber harvests had a significant impact on the natural structure of southern taiga. Until the XX century, the forestry of this region was erratic and the volume of forest consumption was not counted, which led to the destruction of coniferous forests (Dubuc, 1912). The destruction of the taiga took place in the direction from the north-east to the south-west and depended on the degree of economic development of the territory. Peak logging fell on the 50-70 years of XX century. Mighty pine forests were replaced by less productive massifs of secondary pine forests, birch, or pine-birch forests with birch predominance. The share of soft-wooded species has increased to 53%. Species composition of forest cover has changed significantly. At present, the edificatory species are birch, spruce and pine in a ratio of 41.8% to 23.8% to 23.3% (Forest Plan of Kostroma region, 2018). As a result, the share of pine forests in the southern taiga relative to the northern and middle taiga decreased by almost half, and the representation of soft-wooded species has increased many times.

The central part of transect, as compared with the eastern and western regions of southern taiga, belongs to the most densely populated. The relief of its territory is flat or slightly hilly (Zubova, 1990; Borisova et al., 2012). Uplands are separated by vast, often swampy plains. Here the Volga River flows with tributaries.60% of forest lands are characterized by adverse edaphic conditions. 300 years ago, a significant part of the territory (approximately 70%) was covered with coniferous and mixed forests (Atlas of Yaroslavl Region, 1999; Rysin and Savelieva, 2008). The distribution of forests and their species composition was formed depending on the peculiarities of the climate, topography, and soil conditions. Pine, spruce, birch, and aspen were among the main forest-forming species. Pine forests, due to the heterogeneity of forest conditions, were initially unevenly distributed. Highly productive pine forests (I - II bonitet class) grew mainly along river banks and on sandy soils.

This territory was settled a long time ago, with the appearance of the first settlements dating back to the end of the late Paleolithic era (12-13 thousand years BC) (Atlas of Yaroslavl Region, 1999). It had no equal in the development of shipping, trade, and crafts, and was the center of population concentration. The first settlements of Slavs appeared in the VIII-IX century. Since the XI century, the population growth rates have begun to increase rapidly. The mass extermination of coniferous forests began at the same time. Agricultural land was ploughed in large areas, those areas being the best forest land. Quite an important role in the depletion of forest resources belonged to trades related with the forest, from many small handicraftsmen to large sawmills and wood processing companies.

At present, the forest cover area of the central part of the southern taiga is significantly lower than in its eastern part (45.4%) (Forest Plan of Yaroslavl Region, 2018). Forest cover is characterized by "spotting." Now existing forests are mainly the secondary in origin. They are susceptible to diseases, pests, and other pathogenic factors. Their species composition and age structure is strongly disturbed.40% of forest land falls on birch. The predominance of birch takes place everywhere: on sandy terraces, hillsides, floodplains of rivers, and wetlands.

Pine forests are most affected by human activities. The pine share on average in the region is only 7% of forest land. On the sands of Volga right bank it reaches 15%. The average of their age is 70 years, bonitet 2.1 (Forest plan of Yaroslavl region, 2018; Forest plan of Vologda region, 2018).250-300-year-old pines are preserved only as individual trees or on the small sites. The area of separate massifs is 5.5 hectares. Scots pine, as a more competitive species, takes the place of spruce forests on watersheds (Borisova et al., 2012). Qualitative pine undergrowth is observed only on 30% of areas of the cut down forests. Birch dominates with the overgrowing of former agricultural land, while pine occupies only the fourth position (Maslov et al., 2018).

The further from the central part of northern transect, the relief is mostly flat or hilly. In the western area its part are many large and medium rivers (21 rivers), marshes (6% of the area), and lakes (19%, the area is more than 10 km^2). The uplands stretch from southwest to northeast. All rivers belong to the basins of 3 rivers (Volga, Western Dvina, Neva). The total area of the Volga river basin is 59,650 km, or 2/3 of the analyzed region. Forests are the main type of zonal vegetation (Rysin and Savelyeva, 2002; Forest Plan of Tver Region, 2018). Even 300 years ago, this part of the southern taiga was covered with dense forests and vast swamps. The famous Okovsky Forest, which was mentioned in the ancient Chronicles of the XII century, is located in the South-West of transect. The natural pine forest is still preserved around Seliger Lake (Figure 2).

By the end of the 17th century, the forest cover of this region was more than 75.8% (Tsvetkov, 1957). The main share of forest land was occupied by coniferous forests. The development of natural pine and fir forests for centuries occurred in the following sequence. Pine, as a less demanding species for fertility and moisture of soil, was resumed on the places of burnt spruce forest (Dobush, 2013; Deryugin, 2018). Under its canopy the undergrowth of a fir-tree was formed. Until now, these are the most powerful highly productive forest stands.

a   b

Figure 2. Natural Okovsky Forest (Pine Bor) on Klichen Island around Lake Seliger, Tver Region.

The peak of development of the northern territories and the maximum population growth occurred in the XIX century and, especially, at the beginning of XX century. The area of arable land, pastures and hayfields increased at the expense of forests. At the same time, the domestic market for forest consumption began to actively develop, what also led to decrease of their percentage. In 1914, the forest cover fell to 58.1%. Therefore in the second half of the XX century, the main volume of logging and the products of their primary processing were moved to areas of middle and northern taiga.

At present, the forest cover of western territory is 54% (Forest Plan of Tver Region, 2018). Soft-leaved species dominate (56.9%, 33% of which are birch forests). Current southern taiga forests are the secondary forests, in which the species composition is changed. The number of forests with impaired and lost resistance to the effects of diseases and pests is half from their total volume. The area of pine forests here is slightly higher than in the center of transect (19.7%). They are presented as pine forests of different types, but more often they are the mixed spruce-pine and pine-birch forests. Pine is mainly concentrated on elevated places and hills along the Volga and its tributaries. Pine-fir forests with a more complex species composition of the undergrowth (buckthorn, rowan, viburnum, etc.) form on richer soils. A successful pine forest renewal potential and a good quality of pine-tree subgrowth are also observed in this part of the southern taiga (Zakharov, 2010).

Summing up, it should be noted that pine in the forests of the northern and middle taiga, growing in the adverse conditions of species pessimum, is more competitive and dominates over other species of edicators (Kuznetsova and Sautkina, 2019). On the contrary, evident prevalence of birch in the forest cover in the southern taiga (optimum zone of *Pinussylvestris* range) is observed. Pine takes 2-4 positions along the all northern latitudinal transect. The smallest effect of the changing of the species composition is observed in the east of transect. The ratio of pine share in the forest cover from the eastern, central, and western their parts are 1.0: 0.4: 0.5. Pine forests in the central part of transect are the least preserved, which is, apparently, due to the earlier development of this territory by humans. A positive factor is that the natural regeneration of pine, and the quality of pine-tree subgrowth is satisfactory. This indicates that the connection in system "parents-progeny" is not broken.

3. CENTRAL LATITUDINAL TRANSECT: TEMPERATE BROADLEAF AND MIXED FOREST ZONE

The central latitudinal transect passes through center of the Plain, has a 1000 km length, and belongs to the zone of broadleaf and mixed forests. The western slope of the Volga Upland and Volga-Oka interfluve is its beginning. The climate is temperate continental with well-marked seasons of the year: warm and humid summers and snowy winters. There are no sharp fluctuations of edaphic and hydrological conditions within transect. The flat undulating relief of the watersheds alternates with wide valleys. Hydrothermal coefficient varies within 1.6-1.2. The average annual rainfall is 550-600 mm. There are hundreds of large and small rivers, about 300 lakes, and many swamps.

The forests are the main type of zonal vegetation (average productivity is II class of bonitet). 300 years ago they covered most of this territory. Their species composition was formed in the post-glacial period under the influence of climate, relief, and hydrological regimes (Mayevsky, 2006; Romanov, 2008). Forest vegetation was represented by broad-leaved species, spruce, and pine. Broad-leaved forests are mainly associated with river floodplains (Romanov, 2013). Pine and spruce in forests can be both together and separate. They are also part of mixed forests with broad-leaved and soft-leaved species.

Pine occupied mainly sandy soils or swampy soils. Pine massifs earlier occupied the largest territories: on the border with the taiga zone - a pure mono dominant pine forests; in the center and on the border with the temperate steppe zone - mixed broad-leaved–coniferous forests. Compared with the taiga zone, the mixed forests were characterized by a large variety of species composition. Accompanying species were birch, linden, mountain ash, aspen, juniper, etc. These forests have the otherwise same composition and structure (Figure 3a). In the Nature Reserve conditions, trees of birch and pine in mixed pine-birch forests have the same height (bonitet class of both species is I and Ia), and together constitute the forest's upper canopy. In the east, there were mostly spruce forests and broad-leaved forests with local massifs of pine. Broad-leaved (coniferous) forests prevailed in the north-west region. Pine was found only in the form of isolated islands.

Human economic activity during the XIX and XX centuries significantly changed the species and age composition of forests. Forest cover of the territory decreased to 50%.The average age of forests is 46 years. The average bonitet class is 1.4 (Forest Plan of Vladimir Region, 2018). At the present time, large pine forests are preserved only on the border with the taiga zone (territory of Meshchora Lowland, 100-120 m). Mixed pine-birch forests now occupy the largest area. The ratio between these species is closing progressively and change in the direction from north to south of this transect: in the north: pine - 45%; birch - 35% (Forest Plan of the Vladimir Region, 2019); in the south: pine 29%, birch - 37%; (Forest Plan of the Ryazan Region, 2018).

It is noteworthy that the closer to the center of range, the more strongly the interspecies competitiveness of Scots pine decreases. At the same time, the quantity and quality of pine-tree subgrowth significantly decreases during natural reforestation. Satisfactory reproduction is noted in the middle taiga in 83% of areas, 68% in the southern taiga, and only 32% in the temperate broad leaf and mixed forest area (Maleev et al., 1998). The pine-tree subgrowth in the pure pine and mixed pine-birch forests is small and is greatly weakened (Figure 3-b). The natural reforestation of birch, on the contrary, occurs very intensively, and it occupies the released space (Figure 3-c). This allowed us to make an important conclusion that the further natural development of forest cover within this region will lead to an irreversible change of mixed pine-birch forests to the pure birch forests.

Figure 3. Mixed pine-birch forest of Mordovia State Reserve (a); the natural reforestation of pine and birch in the east (b, c) and pine in the west (d) of the central transect of East European (Russian) Plain.

The Moscow region is located in the center of the analyzed transect and the East European Plain. Until the end of the XVII century, 48% of its land was covered with broad-leaved and coniferous forests (Tsvetkov, 1957; Korotkov et al., 2015). Pine, spruce, birch, and aspen were its main edificatory species. Pine massifs grew in the west, on the marshy lands of the Meshchora Lowland. Mixed pine forests were located on the terraces of the Moscow River. As a rule, the plain forests of pine and spruce have a simple mono dominant structure, in the form of pure pine and spruce forests. A large territory was occupied by broadleaf forests (oak, linden, maple, ash) with a more complex structure (Popadiuk et al., 1994; Annenskaya et al., 1997; Rysin, 2012). Currently, the natural forests here are almost not preserved.

Beginning from the XIII century, the region began to lose the natural forests and optimal relationship between edificatory species (Rachelin, 1997; Korotkov et al., 2015; Forest Plan of Moscow Region, 2018). The rapid concentration of population, as well as the development of agriculture and handicrafts led to the forest destruction over a large territory. In the XIV-XVII centuries, when Moscow was the capital of Russia, the area of farmland increased significantly. Slash-and-burn forest-field agriculture turned out to be the most destructive, since mostly rich lands were plowed open. At the same time, the pottery, metallurgy, and other industries began to develop. As consequence, its forest cover at the end of XVII century decreased to 38%.The timberland almost completely disappeared, and the edificatory species were replaced by soft-leaved ones, mainly by the birch (Korotkov et al., 2015). In the XVIII century, the forest area continued to decline and reached a critical level (21%) at the beginning of the XX century, at which the natural landscape for agriculture was completely exhausted.

In the middle of the XX century, the forest cover of the Moscow Region as the result of artificial reforestation almost doubled. A rather high percentage of coniferous in it (40–50%) is supported due to pine and spruce artificial forest plantations (Atlas of Yaroslavl Region ..., 1999). These forests in comparison with the initial ones have a more depleted floristic composition. Most of the existing forests are secondary in origin.

Their species composition and age structure are greatly changed, but the relationship with the zonal ecotopic structure is still maintained (Tikhonova, 2006). The ratio of birch and pine is 39.5% to 20.0% (Forest Plan of Moscow Region, 2018). Forests, which spontaneously arose in the former farmland, are unproductive, fragmented, and weakened. Birch among them also occupies a dominant position.

The most alarming sign is the lack of viable pine progeny during natural reforestation (Pismenny, 1970; Rysin and Savelieva, 1994).The change of species composition is everywhere, including in the reserve conditions. During monitoring of 180-year-old natural pine forests, it was found that the number of trees in the sample plots decreased by 15% over 50 years (Rysin and Savelieva, 1994). Pine-tree subgrowth in them is not numerous. It is strongly weakened, characterized by dwarfism, low resistance to diseases and pests, and dies within 3 years. Lowered vital potential of pine in comparison to birch appears in the form of a more weak interspecific competitiveness. In young mixed pine-birch forests, a birch has a competitive advantage and, as a rule, suppresses a pine. Change of pine to birch occurs at 40-70% of pine forest felling (Pismenny, 1970). The above signs indicate the instability of pine forests in central part of species range (Rysin, 2006).

Further to the west, the latitudinal belt passes through north-western margin of the Central Russian upland. Here the relief of the Plain changes. It is dissected by ravines, gullies, and river valleys (Bitkov, 1998).The natural area consists of two subzones: coniferous-broadleaf and broadleaf forests. Almost half of the land is under the forest. Natural forests carried the imprint of climatic and soil conditions and were characterized by the stability and diversity of tree species. In the main, they were located on watersheds, in types of forest growing conditions C_2 - C_3. This type of forest lands is most prevalent in this territory (70% lands). The predominant species were pine and spruce. In the past, the pine forests and complex spruce forests (with linden) among coniferous-broadleaf forests were distinguished as the most stable and productive forests (Kashkarov, 1908). They also coincided with the watersheds, where the most intensive extermination occurred.

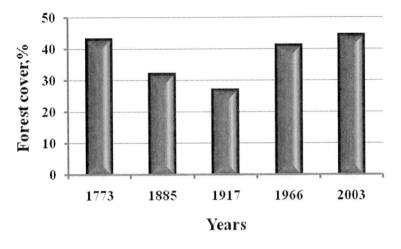

Figure 4. Changes in the forest cover of Kaluga Region for the period from 1773 to 2003. (By Bitco, 1998).

Figure 4 shows the dynamics of change in the area of plain forests for 200 years with the example of the Kaluga Region. In the first half of the XX century, the minimum level of forest cover was 27%. The existing and natural levels are the same. It should be noted that such an algorithm for changing forest area is typical for all another analyzed regions from the zone of temperate broadleaf and mixed forests.

Current forests bear the imprint of the influence of various anthropogenic factors. Many of them are the secondary by origin. The average forest cover is 45.3%, but its forests are unevenly distributed. The maximal level is characteristic for the northern regions – (57.9%), while the minimal for the southern ones is13.5%.Birch dominates among edificatory species, which accounts for half of the forest lands (49.7%). Birch forests are characterized by an affluence of species composition. Pine, after birch and spruce, occupies the 3rd position (9.6%). The area of pine forests is reduced due to the fact that, firstly, the natural regeneration of spruce under the pine comes with the change of pine forests to spruce ones. Secondly, the felling and former farmland are overgrown mainly with soft-leaved species (Aleinikov, 2005; Reshetnikova et al., 2010).

An analogous situation occurs in the west of this transect. The main singularity feature of its landscape is the Smolensk-Moscow Upland (320 m), which occupies almost half of the territory. The relief is wavy with a great number of hills, lakes, ravines, and rivers with deep enough incised (up to 50 m) river valleys. Initially, highly productive coniferous, broadleaved, and mixed forests grew here. Most mighty of them is the Okovsky Forest (Alekseev, 2006). In ancient times, it was formed on the Plain between Tver, Smolensk, and Novgorod, where glaciation did not reach (Figure 2).

The extermination of natural pine forests began in the XVIII-XX centuries, but their area decreased most of all in the first half of XX century (Pozdeev, 2000). At the present time, most of them have been replaced by secondary forests. As a rule, their formation took place with the change of species. The forest cover of the west territory is 42% (Forest Plan of Smolensk Region, 2018). Birch holds the first position among edificatory species (46.2%). Pine moved to 4th place (7.4%) after spruce and aspen. Coniferous forests grow in the north and north-west of the region, and soft-leaved forests grow everywhere (Shkalikov, 2004). The natural pine regeneration occurs unsatisfactory due to the poor quality of pine-tree subgrowth (Osipova, 2016) (Figure 3d). Weakened progeny die within 3-5 years.

The protracted systemic crisis of the 1990's century improved the ecological situation in Russia because of a significant decline in production in the heavy and chemical industries of the country by 50-70% compared to the 1980's.Inthe same years, the volume of silvicultural works dropped sharply. As consequence, the mechanisms for regulating the species composition were given to Nature itself, and it disposed of the opportunity provided in its own way. In turn, we received valuable factual material regarding the spontaneous development of forest vegetation in the center of the ranges of two edificatory species. According to the data, Scots pine, as a less competitive species, began to be replaced by the birch from the liberated forest lands. Pine is also not renewed on the former farmland. Its replacement with soft-leaved species comes at a very high rate (1-2% every 5 years) (Kuznetsova, 2010; Kuznetsova and Sautkina, 2019). In the future, this will lead to a break of the range of pine in half. The Greenpeace "Forests of Russia" map confirms this negative prediction (Figure 5). The map clearly shows that the zone where the dominance of conifers has changed to the soft-leaved passes along the latitude of the central part for most of the Eurasian continent. The formed gap of the range of pine in half on the East European Plain also fully coincides with this zone.

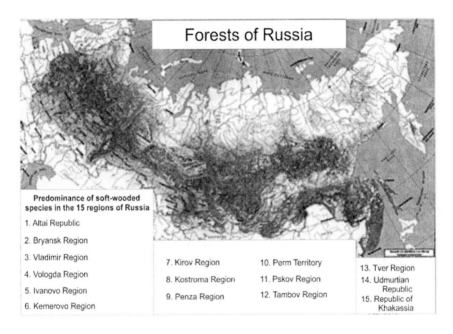

Figure 5. The Greenpeace map "Forests of Russia" (by Peskov, 2005).

The analysis carried out in this section of the Chapter shows that the forest cover of territory of the central transect, which passes through the center of East European Plain and the range center of *Pinus sylvestris*, is, like 300 years ago, 40-50%.At the same time, the species composition of the forests has undergone a significant transformation. The birch began to dominate over the pine (except in the northeastern part of the transect). There is a stable loss of pine forests. Profound changes in sexual reproduction occurred, the consequence of which was a violation of connection in the "parents-progeny" system and a decrease in the quality of seed progeny of pine (Kuznetsova and Sautkina, 2019).Greenpeace data and our analysis indicate that replacing coniferous to soft-leaved forests is a problem of gene pool of the *Pinus sylvestris* L., and not only the territory in which this species grows. In Russia, this process affected regions that are located in the center and less on the periphery of range. The reasons for instability of pine forests are not known and require further study.

3. Southern Latitudinal Transect: Temperate Steppe Zone

The southern latitudinal transect passes through temperate steppe zone of the East European Plain. The climate is moderately continental with insufficient and unstable moistening. Its continentality and the intensity of solar radiation from east to west weaken due to the decrease of the influence of the Atlantic Ocean and with the western transfer of air masses (Kostowskay and Stulishapka, 2014). The density of the river network is three times lower than the average level in Russia (Kurdov, 1975). The average annual rainfall is 450-550 mm. The hydrothermal coefficient goes over via 1. Droughts occur with a frequency one every of 3-4 years (Svintsova et al., 2014; Kuznetsova, 2015). In the east of transect, powerful chernozems (humus content 9%) are deposited, the formation of which was promoted by a warm climate and steppe vegetation (Popov, 2003). To the west, they change over into predominantly gray and chernozem soils. Relief consists of heavily branched beams and ravines, often not covered with forest vegetation, elevated and low-lying sites, and wide valleys (Neustruev, 1930).

The initial forest distribution corresponded to the natural structure of the landscape (Milkov, 1977).Forests were usually placed at elevated sites and to the right of banks of rivers. The main forest-forming species are pine, oak, and birch. The steppe occupied mainly flat sites of interfluvial areas and terraces of plains (Milkov, 1954; 1966). Dynamic equilibrium between a forest and a steppe has been formed during thousands of years (Novoseltsev et al., 1985). The temperate steppe (forest-steppe region) has always been characterized by an island type of forest distribution, which is confirmed by paleontological data (European Paleontology, 1982). As the continentality of climate decreases (from east to west), the diversity of forests increases and their species composition becomes more complex. Forests of the western part of transect are characterized by the greatest floristic richness. The forest-steppe is a zone of productive forests (I – III bonitet) and highly productive pine forests (Samofal, 1925), and the highest V zone in pine seed quality (Kuznetsova, 2015; Kuznetsova and Sautkina, 2019). A comparative study of woody species used for forest belts revealed once more advantage of the *Pinacea* species. For example, a pine in terms of carbon sequestration is a 30% more efficient species compared to poplar (Huitao et al., 2014).

Fresh bors in the analyzed territory occupy flat sites (second above floodplain terraces) or gentle slopes (Development of scenarios ..., 2017).They are represented by pure plantings of pine of the II bonitet class mixed with birch and aspen. Pine in them is characterized of high quality wood. Fresh simple subor are relatively poor habitats. The relief is usually quite gentle and slightly wavy. The two-tier structure of indigenous types of pine forests is typical for them (I – Iabonitet class).The first tier is formed mainly by pine, the second by the predominance of oak. Fresh complicated subor occupy fertile habitats formed on loamy sand and sandy soils. Pine forests have the complex structure. Pine prevails in the first tier. In the second oak, linden, and maple prevail. In the third linden, apple, and pear prevail. They are characterized by high productivity (Ia –Ibbonitet). The undergrowth consists of hazel, spruce, raspberry, rowan, and other shrubs.

In forest-steppe, the environment-forming function of forests comes in first place due to the low forest cover of the territory, high ploughing of lands (approximately75%), and intensive carbon deposition effect (unlike agrocenoses). In the 1960's to1980's, the artificial reforestation in the forest-steppe region was carried out mainly through the creation of artificial forest plantations and forest belts. Pine forests of forest-steppe are mainly represented by monocultures. Most of them are characterized by good quality and stability (Zelenin, 2000). Their average preservation of pine plantation in the further tending of them is 60-85%. At the same time, a pine without tending is displaced by birch or aspen (Koldanov, 1966).

Today, the main share of forest plantations (more than 80%) in Russia are created by means of the landing. Landing method can be in the form of rows, bands, corridors, biogroups, or randomly way (Martynov et al., 2008). The main technology is the furrowed method with landing seedlings in the bottom of furrow. It should be noted that the formation of the forest environment begins from zero when crops are created in rows with solid tillage and requires more time and cost. In a result of conducted research (2012–2016), the method of creating pine plantations by the biogroups turned out to be more efficient (Semenov and Kharchenko, 2017). It has been shown that the surface roots of the aspen within the biogroup are shifted in the shaft. The dense landing of planting material leads to an accelerated connects of crowns, which gives rise to the transition of this area into woodland area. Intraspecific selection inside the biogroups is effectively carried out. The overall appearance of the biogroup takes the form of a "pyramid." The connection of leading trees in the future creates a main tree canopy while maintaining the fond of the forest environment corresponding to a particular type of forest growing conditions. The forest environment thus formed is as close as possible to the natural ones without the use of costly silvicultural measures.

The southern latitudinal transect begins from central part of the Oka-Don Lowland and its length is 630 km. The relief is weakly undulating plain with many rivers, valleys, and ravines. The eastern part of transect is characterized by powerful chernozem soils, which were formed in the post-glacial period (Neustruev, 1930). Vegetation cover is represented by forests, steppes, and meadows, and includes 1240 plant species. 300 years ago, forests occupied 40.5% of the territory. Their main type is Pine Bor.

At present time, 2/3 of the territory of the Oka-Don Lowland is plowed, and almost all farmland is located on black-earth soils. The steppe vegetative communities suffered more. Initially, the steppe vegetative communities suffered to a greater extent and the forest ones to a lesser degree due to low productivity soils. By the end of the XVIII century, the forest area decreased to 16%. The forest was used as a source of building material, fuel, and charcoal. Over the next 150 years, the forest cover decreased threefold relative to the optimal level and now amounts to 11.7%

(Tambov Forest, 2006). Most of the current forests have the secondary origin. They differ from natural forests in age and diversity.

Figure 6. Natural pine-tree subgrow thin Tsninsky Bor, Tambov Region.

Today, the share of Scots pine is 43.1% from the total area of forests in the eastern part of analyzed region. 80% of all works by the artificial reforestation fell on pine (Arsyukov et al., 1973). The seed sources were the Tsninsky and Ilovaiskypine forests. A cytogenetic study showed that the genetic material of autochthonous maternal forests and the forest plantations created from their seeds are not significantly different (Ermolaeva, 2009).

Pine is the dominant species in the eastern part of southern transect, with pine comprising 46.1%, birch 17.2%, and oak 15.7% (Forest Plan of Tambov Region, 2018). However, the share of soft-wooded species has steadily increased (Tambov Forest, 2006). Most of the pine resources are concentrated in Tsninsky Bor and Ilovaisky Bor. The remaining massifs are represented by small islands, forest belts, and artificial plantations. In contrast to the central transect, here a large number of viable pine undergrowth is formed (Figure 6).

The Oka-Don Lowland further along transect passes into the central part of temperate steppe zone, the main relief structures of which are the Oka-Don Lowland and the Kalachskaya and Central Russian Highlands. The river Don flows between these Lowlands. The dominant soils are chernozems, which make up 3/4thsofthe region's area (Akhtyrtsev and Sushkov, 1983).

Until the XIX century, this territory remained the "Wild Field" (Akhtyrtsev et al., 2006). The floodplains of rivers were covered in powerful forests. The natural level of forest cover was 31%.Forest extermination had a long history here (Anichkina, 2015). Since ancient times, the local population engaged in smelting metal. Charcoal for stoves was procured from nearby pine and oak forests. The first state-owned factories, for which the forests were massively cut down, also appeared here. The river Don has always been a trading link between the northern and southern regions of Russia. In the XIII-XVI centuries, the construction of ships was carried out in its shipyards. The area of floodplain forests has decreased by 3-5 times. In the beginning of the XX century, forest cover reached a minimum (6.2%) as a result of population growth, ploughing of lands, and industrial growth (Musievsky, 2013).

At present, the forests of this region are represented as small massifs. The largest of them is Usmansky Bor (Sinitsin, 1982). It is located between the rivers Voronezh and Usman, and is a natural standard of the Central Russian forest-steppe. In 1723, Usmansky Bor was declared a reserve territory. Several types of pine forests are located in it. The most productive is the massifs with Iabonitet, which grow on sandy soils with close depth of groundwater (Figure 7a). The pine trees at the age of 70 reach a height of 29 m. Here, we carry out our monitoring research on Scots pine (Kuznetsova, 1996, 2015, 2019, etc.). The second type is a pine forest, which is located on the sands of dunes with far-lying groundwater (terrace on the left bank of Voronezh River). Here the pine forests have I bonitet.

At present, the forest cover of the central part of transect is7.2% in the north (Forest Plan of Lipetsk Region, 2018) and 8.1% in the south (Forest Plan of Voronezh Region, 2018). Most forests have the protective status. In the north, pine among edificators is the dominant species, (34.2%), followed by oak (28.8%). To the south, this species representation occupies the second position (29.1%). Birch in this list takes the third place (12.9%). A large contribution to the restoration of optimal ratio between these species had been made the silvicultural work during the 1960's to 1980's. As rule, a sufficient amount of viable pine-tree subgrowth is formed near the forest wall and inside the sparse pine forests. Former farmlands are overgrown mainly with pine and/or birch (Figure 7b).

a   b

Figure 7. Umansky Bor (a); pine regeneration on former farmland (b), Lipetsk Region.

The western part of the southern latitudinal transect is located on the southwestern slope of the Central Russian Upland. Here is the upper basin of the Dnieper and Don rivers. Chernozem soils occupy ¾ths of this territory. According to the geobotanical classification, this is the Kursk district of oak forests and sod-mixed herbs steppes of the Central Russian forest-steppe province (Alekhin, 1924). It is characterized by the zonal (upland oak and steppe meadows) and extra-zonal (meadows, shrubs, phytocenoses of chalky outcrops) types of vegetation cover. Oakwoods are usually associated with rivers, beams, ravines, and marshes. The plains, as a rule, are occupied by meadow steppes. The share of pine forests is small.

Figure 8. Natural pine forest on the slope of the Cretaceous sediments, Kursk Region.

Undisturbed forest-steppe presents the alternation of steppe vegetation with oak forests growing along river valleys, along the slopes of beams and watershed plateau. One of the key components of its stability is the optimal ratio between the steppe and the forest, as well as between species-edificators. During the postglacial-holocene, the species composition of the forests in the territory of the Central Russian Upland was changing with a frequency of 2.5-1.0 thousand years. Its last formation took place 1-2 thousand years ago (Smirnova et al., 2001). Oak after the Ice Age became the dominant species and has occupied, and still occupies, most of the forest land (Chendev, 2008; Chendev et al., 2008). Pine forests were pushed to the sands and chalk sediments. Their descendants still grow on poor unsuitable lands (Figure 8). These natural forests are characterized by low height (III-IV class of bonitet) and different stand density.

A significant impact on the current state of forests is imposed human economic activity. The initial forest cover of the forest-steppe in the West of Russia was almost half of its territory (Degtyar et al., 2016). Active settlement of the region began in the XVI - XVII centuries (Virsky, 1925). During the last 300 years, the forest area has been reduced by three times (up to 16%) as a result of intensive deforestation to increase the area for farmlands and for the needs of the mining industry. The integrity of initially uniform woodland was disturbed. According to Tsvetkov data (1957), 2/3rds of the natural forest-steppe forests had disappeared by 1914. Their species composition has undergone significant transformation. Most of the pine forests and almost all Cretaceous pine were destroyed in the XVIII century. The area of pine forests had decreased from the North-East of the Central Russian upland (Terekhin, 1989). In 1927, the forest area decreased to 5%, and the formerly forest-rich region became treeless. The result of the destruction of forests was an increase in the area of the gully-beam network by 15%, the strengthening of erosion processes, and the limit of the lower border of the optimum of zonal water protection forest cover was attained (Kuzmenko et al., 2013).

Now, the forest area is about 9% due to the large amount of forestry work carried out in the 1950's to 1980's in the territory of the Kursk and Belgorod Regions (Forest plan of the Kursk Region, 2018; Forest plan of the Belgorod Region, 2018). Artificial plantations of pine and oak grow on ¼th of the forest lands. Oak occupies a large part of its area (approximately 80%).The share of pine is significantly lower (10.1-11.8%). Almost all present pine forests are the derivatives. Their living state and productivity are lower than in the eastern and central parts of southern transect. The natural reforestation of pine is abundant on the edges and open sites of slopes, as well as in sparse pine forests (0.4-0.5 stand density) and on the spaces after fires. The shares of self-sowing of birch and pine on the spaces after fires are 1: 1, or 1: 2, accordingly (Ushatin and Mamonov, 2012). In 60-year-old pine cultures, the viable pine-tree subgrow thunder the forest canopy is absent.

The processes described in this section indicate that the natural forest cover of the Central Russian forest-steppe during the economic development of the territory of the temperate steppe zone has undergone significant quantitative and qualitative changes, and now it is almost completely the antropogenic. Pine, oak, and birch are the main forest-forming species. They are represented by forest plantings of I-III classes of bonitet, characterized by a good and satisfactory state and a relatively high level of biological diversity (Semenov et al., 2017b). The main reasons for reducing of pine forests are perennial intensive anthropogenic pressure, forest exploitation, and forest fires. Despite this, the area of pine forests has not changed significantly. Table 3 shows the dynamics of areas of the forest-covered lands occupied by pine trees in 4 studied regions of the southern transect over 20 years. They, as of 01 January 2017, have not changed significantly in the Kursk and Belgorod Regions, have increased in the Kursk and Tambov Regions, and have decreased in the Voronezh Region (Development of scenarios..., 2017).

The representation of pine and birch, if we follow by the latitudinal vector in the direction from east to west, is significantly reduced, with pine reducing from 46.1% to 31.6% to 10.1% and birch from 17.2% to 11.7% to 4.6%. Another distinctive feature is the shift at the dominance of pine to oak within the southern transect. If in the east the ratio of pine to the oak is 1: 3, then in the west, on the contrary, there is an 8-fold excess of oak to the pine. The process of natural regeneration of pine is sufficient in quantity and quality of seed progeny. The data of the monitoring studies showed that the pine of the temperate steppe zone is characterized by abundant annual fruiting, rapid growth, drought resistance, qualitative seed progeny, etc. (Kuznetsova, 1996, 2015, 2019; Kuznetsova and Sautkina, 2019). Therefore, a pine forest, presumably, in conditions of global climate change will be different stability, but only to a certain limit. At the same time, the increase in the impact of secondary negative factors (forest fires, pests, and diseases) are not excluded (Development of scenarios ..., 2017).

Table 3. Dynamics of areas (forest-covered lands) occupied by Scots pine in 1997, 2013 and 2017 years

Russian Federation Subject	Areas of forest-covered lands occupied by Scots pine, thousand hectares		
	1997	2013	2017
Tambov Region	141.3	154.2	152.9
Lipetsk Region	53.7	53.0	54.9
Voronezh Region	103.5	86.7	86.5
Kursk Region	22.2	25.9	26
Belgorod Region	19.2	20.1	19.7

CONCLUSION

The analysis shows that during the twentieth century, birch is increasing its volume almost throughout the East European Plain. Pine, on the contrary, began to lose its positions in the regions in which it traditional grown. Interspecific species competitiveness has decreased. The quantity and viability of pine-tree subgrowth during natural reforestation has become much lower. The zone of greatest population instability passes through the center of range (along latitude). This indicates that a widening gap had emerged between the northern and the southern half of continuous range of pine. Now living pine forests still correspond to the genotypic and phenotypic species norm and have a holistic and continual gene pool. However, the next generation of forest, which arises during sexual reproduction, is weak and non-viable.

A review of experimental data showed that Scots pine has violations in the "parents-progeny" system, which are most pronounced in the central part of its range. This phenomenon may be caused by several reasons, the main of which are the evolution of species and forest cover on the whole, global climate change, the degree of land deforestation, and anthropogenic stress. Therefore, a question arises as to which of these reasons was decisive and has resulted in the loss of the ability of the species to form viable seed progeny. We have reason to believe that anthropogenic stress, deforestation, and climate change may accelerate the change of species composition, but are not its original causes. So, by the beginning of the XX century, 2/3rds of the forests on the territory of the forest-steppe zone were cut down. Disturbances of its bioclimatic system are accelerating due to a global climate change, a drop in the groundwater level, soil erosion, and disturbance of soil microflora (Cheverdin et al., 2017; Sautkina, 2017; Sautkina and Kuznetsova, 2018). Nevertheless, the pine on the territory of forest-steppe retained the ability to form viable seed progeny, and normal reproductive capacity in the center of the range it has lost. This may serve as an argument that the reasons of destruction of the "parents-descendants" connection are deeper. Apparently, they are related to the evolution of forest cover and the change of dominants caused with evolutionary processes.

The forest cover of the East European Plain has been developing for over a hundred years in the direction of changing the species composition. This process in the present historical stage affected only the central part of the range. Birch on the territory of rupture displaces pine and stably retains own leadership. *Pinus sylvestris*, the previous main edificator of temperate broadleaf and mixed forests in the centre of Plain, has already ceded leadership to birch, and, perhaps, is irreversible. On the Eurasian continent, this process captured fairly big patch in the central part of range from east to west, as can be clearly seen on the map "Forests of Russia" (Peskov, 2005). At the current high rate of alteration of the species composition (1-2% at 5 years), this will lead to the space separation of the northern and southern parts of the range, the change of geography of the pine dissemination, and the loss of biological diversity of the plain forests.

REFERENCES

Abaturov, A. V. (1995). Influence of forest use on the species diversity of forest vegetation of the East European plain. *Biological diversity of forest ecosystems.* MGUL, Moscow: 227-229.

Abaturova, M. P. and Khromova, L. V. (1984). Factors that ensure the formation of pine populations in the swamp. *Features of formation of Scots pine populations.* Nauka. Moscow. 56-64.

Akhtyrtsev, B. P. and Sushkov, V. D. (1983). *The condition of the forest in Lipetsk region.* Publ. VSU. Voronezh.

Akhtyrtsev, B. P., Akhtyrtsev, A. B. and Yablonskikh, L. A. (2006).Soils of the Voronezh region. *Proceedings of VSU. Series: Chemistry. Biology. Pharmacy.* 1: 85-95.

Aleynikov, O. I. (2005). *Forest resources of Kaluga region. Issues of archeology, history, culture and nature of the Upper Reaches.* Publ. "Polygraph-Inform." Kaluga.

Alekseev, L. V. (2006). *Western lands of pre-Mongol Russia.* Nauka. Moscow.

Alekhin, V. V. (1924). Zonal and extrazonal vegetation of Kursk province in connection with the division into natural areas. *Eurasian Soil Science.* 1-2: 71-78.

Anachkina, N. B. (2015). The state of the forests of the Lipetsk region as a result of the interaction of nature and man. *Advances in current natural sciences.* 12: 64-67.

Annenskaya, G. N., Zhuchkova, V. K. and Kalinina, V. R., etc. (1997). *Landscapes of the Moscow region and their current state.* Smolensk humanitarian University Publ. Smolensk.

Arsukov, P. A., Bugaev, V. A. and Dudaev, S. M. (1973). *Experience in forestry in the Tambov Region.* Publ. VSU. Voronezh.

Atlas of Yaroslavl Region. Geography. History. (1999). Publ. DICK. Yaroslavl.

Artyomov, V. A. (1985). Microfenology of the male generative cycle of pine and spruce. *Integrated biocenological studies of coniferous forests of the European Northeast.* Syktyvkar: 56-69.

Bagaev, S. S. and Chudetsky, A. I. (2018). Results of thinning in deciduous-spruce plantations of the Kostroma Region. *Forestry information.* 1: 5-20.

Bitkov, L. M. (1998). *Forestry of the Kaluga Region.* Zolotaya Alleya. Kaluga.

Bobkova, K. S. (1987). *Biological efficiency of coniferous forests of the European Northeast.* Nauka. Leningrad.

Bobrov, E. G. (1978). *Forest-forming conifers of the USSR.* Nauka. Leningrad.

Borisova, M. A., Bogachev, V. V. and Maracaev, O. A. (2012). The forest formations of Western floristic district of the Yaroslavl Region. *Izvestia of Samara Scientific Center of the Russian Academy of Sciences.* 14(1): 974-977.

Bulygin, N. E. and Yaroshenko, V. T. (2001). *Dendrology.* Nextbook. 2nd Ed. MGUL Publ. Moscow.

Butorina A. K., Kalaev, A. N. and Mironov, A. N. (2001). Cytogenetic variability in Scots pine populations. *Russian Journal of Ecology*. 3: 206-210.

Chendev, Y. G. (2008). *Evolution of forest-steppe soils of the Central Russian upland in the Holocene*. GEOS. Moscow.

Chendev, Yu. G., Petin, A.N., Serikova, E. V. and Kramchaninov, N. N. (2008). Degradation of geosystems of the Belgorod Region as a result of economic activity. *Geography and Natural Resources*. 4: 69-75.

Cheverdin Yu. I., Vavin V. S. and Ahtyamov A. G. Influence of melioration methods to the growth of woody species. *Forestry Bulletin*. 21 (6): 13-19.

Degtyar, A. V. Grigorieva, O. I. and Tatarintsev R. Yu. (2016). *Ecology of Belogoriya in numbers*. Publ. "CONSTANTA." Belgorod.

Deryugin, A. A. (2018). Features of population growth under the canopy of birch forests in the southern taiga of the Russian Plain. *Forestry information*. 1: 21-30.

Development of scenarios for adapting of the forestry system in managed forests of the forest-steppe region and the steppe region of the European part of Russian Federation in connection with the expected climate changes (2017).VNIILGI Sbiotekh. Voronezh.

Dobush, I. M. (2015). *Problems of forest biodiversity conservation in Tver Region*. TOUNB named after A. M. Gorky. Tver.

Dubuc, E. F. (1912). Forests of the Kostroma region in the natural and historical relation (the General characteristic). *Materials for evaluation of lands of Kostroma Region*. Kostroma. 13(1): 1-102.

Dudin, V. A. (2000). *History of the forests of Kostroma*. DiAr, Kostroma.

Dudin, V. A. (2011). *Forests of the land of Kostroma*. Line Graph Kostroma. Kostroma.

Ermolaeva, O. V. (2009). *Cytogenetic assessment of the condition of Scots pine plantations of Tsninsky, Usmansky Bor and some urban systems (using the example of the city of Voronezh)*. Publ. VSU. Voronezh.

Filipchuk, A. N. and Moiseev, B. N. (2001). Contribution of Russian forests to the carbon balance of the Planet. *Russian Journal of Forest Science*.5: 8-23.

Forest plan of the Belgorod Region (2018). URL: http://docs.cntd.ru/document/469023824.

Forest plan of the Voronezh Region (2018). URL: http://docs.cntd.ru/document/469705740.

Forest plan of the Kostroma Region (2018). URL: http://dlh44.ru/i/u/NPA/les_plan/pril_ch1_ch2.pdf.

Forest plan of Kursk Region (2018). URL: https://adm.rkursk.ru/index.php?id=138&mat_id=585&sort_field=76&sort_order=0.

Forest plan of the Lipetsk Region (2018). URL: http://extwprlegs1.fao.org/docs/pdf/rus155962.pdf.

Forest plan of the Moscow Region (2018). URL: http://old.klh.mosreg.ru/wood_plan/2273.html.

Forest plan of the Ryazan Region (2018). URL: http://hcvf.ru/ru/regions/ryazanskaya-oblast.

Forest plan of the Smolensk Region (2018). URL: http://les.admin-smolensk.ru/files/198/lesnoy-plan-smol-obl-2016.pdf.

Forest plan of the Tambov Region (2018). URL: http://les.tmbreg.ru/files/Lesnoy%20plan/Proekt_lesnogo_plana_Tambovskoy_oblasti.pdf.

Forest plan of the Tver Region (2018). URL: http: //минлес.тверская область.рф/npbaza/npdokumentylesnoyplanreglamentyLesnoj_plan_Tverskoj_oblasti_po_191pg_(1pg_Prilozhenie).pdf.

Forest plan of the Vladimir Region (2018). URL: http://extwprlegs1.fao.org/docs/pdf/rus155957(1).pdf.

Forest plan of the Vologda Region (2018). URL: https://vologda-oblast.ru/dokumenty/territorialnoe_planirovanie/lesnoy_plan_vologodskoy_oblasti/1715287/.

Forest plan of the Yaroslavl Region (2018). URL: http://docs.cntd.ru/document/934030152.

Glotov, N. V. (1983). Assessment of genetic heterogeneity of populations: quantitative characteristics. *Russian Journal of Ecology*. 1: 3-10.

Harfouche, A., Meilan, R. and Altman, A. (2014). Molecular and physiological responses to abiotic stress in forest trees and their relevance to tree improvement. *Tree Physiology*. 34(11): 1181-1198.

Huitao, SH., Zhang W., Yang, X., Cao J., etc. (2014). *Carbon storange capacity of different plantation types under sandstorm source control program in Hebei Province, China.* Chinese Geographical Science. 24(4): 454-460,

Isaev, A. S., Nosova, L. M. and Puzachenko, Yu. G. Biological diversity of Russian forests. (1995). *Biological diversity of forest ecosystems.* MGUL, Moscow: 40-43.

Isakov, Yu. N., Avramenko, R. S., Butorina, A. K., Ievlev, V. V., Kuznetsova, N. F., etc. (1989). System of seed reproduction of wood plants and breeding. *Forest genetics, selection and physiology of wood plants.* Chief editor. S. A. Petrov. Moscow: 47-54.

Kashkarov, V. M. (1908). *Geographical sketch of the Kaluga province.* Kaluga.

Koldanov, V. Ya. (1966). *Change of species and reforestation.* Lesnayapromyshlennost. Moscow.

Korotkov, S. A., Kisilev, V. V., Storozhenko, L. V., etc. (2015). About the directions of the forest formation process in the North-Eastern Moscow region. *Forestry engineering journal.* 3: 41-54.

Kostovska, S. K. and Stulyshapku, V. O. (2014).Climatic conditions of the Central Russian forest-steppe in the second half of XIX – first half of XX century *Tomsk State University Journal.* 19(1): 234-241.

Kucherov, I. B. and Zverev, A. A. (2012). Lichen pine forests of middle and northern taiga in European Russia. *UT Research Journal. Biology.* 3 (19): 46-80.

Kurdov, A. G. (1975). *Maps of river flow and temporary streams.* Publ. VSU. Voronezh.

Kurnaev, S. F. (1973). *Forest vegetation zoning of the USSR.* Nauka. Moscow.

Kuzmenko, Ya. V., Lisetsky, F. N. and Kirilenko, J. A. (2013). Ensuring optimum water protection of forest cover at the basin organization of nature management. *Izvestia of Samara Scientific Center of the Russian Academy of Sciences.* 15 (3-2): 652-657.

Kuznetsova, N. F. (1996). Incompatibility and stages of its display in Scotch pine. *Russian Journal of Forest Science.* 5: 27-33.

Kuznetsova, N. F. (2009). Effect of climatic conditions on the expression of self-fertility in Scots pine. *Russian Journal of Ecology.* 40(5): 390-395.

Kuznetsova, N. F. (2010). Population instability of coniferous forests and ways to preserve the biodiversity of forest tree species in the Central Federal District. *Principles and methods of biodiversity conservation.* Mari State University. Yoshkar-Ola. 367-370.

Kuznetsova, N. F. (2012). Self-fertility in Scots Pine as system for egulating close relationships and the species survival in adverse environment. *Advances in genetics research.* V. 9. Nova Science Publ., New York, 83–106.

Kuznetsova, N. F. (2015). Development of nonspecific and specific reactions in *Pinus sylvestris L.* at the population level in the stress gradient of dry years. *Russian Journal of Ecology.*5: 332-338.

Kuznetsova, N. F. (2019). *Introduction to the tree-plane corpuscular-wave biology of forest tree species.* Nova Science Publ. New York.

Kuznetsova, N. F. and Sautkina M. Yu. (2019). Forest state and dynamics of their species composition in the Federal Distinct of Russia. *Forestry information.* 2: 25-45.

Levitt, J. (1980). *Responses of plants to environmental stresses. Chilling, freezing and high temperature stress.* Acad. Press. V. 1. New York.

Lvov, P. N., Ipatov, L. F. and Plokhov, A. A. (1980). *Forest formation processes and their regulation in the European North.* Lesnayapromyshlennost. Moscow.

Maleev, K. I., Deryagin, V. T., Shevanyuk, I. L. and Aleksenkov, Yu. M. (1970). Forest cultures and the state of reforestation in the Perm Region. *Lesnoe hozyajstvo.*2: 32-33.

Mamaev, S. A. (1973). *Forms of intraspecific variability of woody plants (on the example of the Pinaceae family in the Urals).* Nauka, Moscow.

Mamaev, S. A. and Makhnev, A. K. (1996). Problems of biological diversity and its maintenance in forest ecosystems. *Russian Journal of Forest Science.*5: 3-10.

Martynov, A. N., Melnikov, E. S., Kovyazin, V. F., Anikin, A. S., Minaev, N. V. andBelyaeva, N. V. (2008). *Basics of taxational economy and forest survey.* Textbook. OOO Publ. "Lan." St. Petersburg.

Maslov, A., Gulbe, A., Gulbe, J., Medvedeva, M. and Sirin, A. (2018). Assessment of the situation with the overgrowing of agricultural land by forest vegetation on the example of the Uglich district of the Yaroslavl region. *Sustainable forestry.* 4: 6-14.

Mayevsky, P. F. (2006). *Flora of the middle zone of the European part of Russia.* 10th ed. KMK, Moscow.

Milkov, F. N. (1954). Terrain types and landscape regions of CChO. *News of the All-Union geographic society.* 86(4): 336-346.

Milkov, F. N. (1966). *Landscape geography and practice issues.* Mysl. Moscow.

Milkov, F. N. (1977). *Natural zones of the USSR.* 2nd ed. Mysl'. Moscow.

Morozov, G. F. (1971). Doctrine of types of plantations. *Selected Works.* Part 2. Nauka. Moscow.

Morozova, O. V. and Tikhonova, E. V. (2012). Differentiation of forest communities in the South-Western part of the Moscow region. *Izvestia of Samara Scientific Center of the Russian Academy of Sciences.* 14(1): 1073-1077.

Neishtadt, M. I. (1964). Features of the development of forests in the USSR in the Holocene. *Modern problems of geography.* Publishing House of the Academy of Sciences of the USSR, Moscow. 207-214.

Nesterov, A. I. and Fedotov, V. I. (2005). To the question of the northern border of the forest-steppe zone of the Central Russian Upland. *Proceedings of Voronezh State University. Series: Geography. Geoecology.* 2: 151-154.

Neustruev, S. S. (1930). *Elements of soil geography.* Selhoziz. Moscow, Leningrad.

Novoseltsev, V. D. and Bugaev, V. A. (1985). *Oakeries.* Agropromizdat. Moscow.

Osipova, N. B. (2016). *Landscape-ecological analysis of the forests of the Smolensk lake area and optimization of forest management.* Immanuel Kant Baltic Federal University Publ. Kaliningrad.

Paleontology of Europe over the past hundred thousand years: an Atlas-monograph. Ed. A. A. Velichko. (1882). Nauka. Moscow.

Pardaeva, E. Yu., Mashkina, O. S., and Kuznetsova, N. F. (2013). State of the generative sphere of Scots pine as a bioindicator of forest stability in the Central Chernozem Region due to global climate change. *Proceedings of the Saint Petersburg Forestry Research Institute.* 2: 16-21.

Peskov, V. (2005). Your card is a bit. *Russian forest newspaper.* 1(79): 2.

Pisarenko, A. I. (1977). *Reforestation.* Lesnayapromyshlennost. Moscow.

Pisarenko, A. I, and Strakhov, V. V. (2004). *Forestry of Russia: from use to management.* Publ. House "Yurisprudenciya." Moscow.

Pisarenko, A. I., and Strakhov, V. V. (2014). About long-term forest changes in the European part of Russia. *Lesnoehozyajstvo.* 3: 2-7.

Pismennyj, N. R. (1970). About the future of coniferous forests of Russia. *Lesnoehozyajstvo.* 1: 54-58.

Pokrovskaya, I. M. (1966). Spores and pollen from paleontological sediments. *Paleontology.* T. 141. Publ. VSEGI. Leningrad: 322-369.

Popadyuk, R. V., Chistyakova, A. A., Chumachenko, S. I., et al. (1994). *East-European broad-leaved forests.* Nauka. Moscow.

Pozdeev, V. P. (2000). *Specially protected natural areas in the context of regional development: textbook.* SSPU. Smolensk.

Popov, V. K. (2003). *Birch forests of the Central forest-steppe of Russia.* Publ. VSU. Voronezh.

Pravdin, L. G. (1964). *Scots pine. Variability, intraspecific taxonomy and selection.* Nauka. Moscow.

Pushkova, L. N. (1968). Moskvoretsko-Okskaya plain in the process of mastering it by man. *Natural resources issues.* 207 (1): 58-80.

Rakhilin, V. K. (1997). Forests of the Moscow Region in the XVII century. *The history of the study, use and protection of natural resources in Moscow and the Moscow Region.* Janus-K. Moscow.

Reshetnikova, N. M., Mayorov, S. R., Skvortsov, A. K., etc. (2010). *Kaluga flora: an annotated list of vascular plants of the Kaluga Region.* KMK. Moscow.

Romanov, V. A. (2008). *Landscapes of the Vladimir Region: a textbook.* Publ. VlGU. Vladimir.

Romanovsky, M. G. (2002). *Productivity, stability and biodiversity of the lowland forests of the European Russia.* Publ. MSFU, Moscow.

Rysin, L. P. (1995). Type of ecosystem as an elementary unit in the assessment of biodiversity at the ecosystem level. *Russian Journal of Ecology.* 4: 259-262.

Rysin, L. P. (2006). The historical factor in the modern succession dynamics of the forests of the center of the Russian Plain. *Russian Journal of Forest Science.* 6: 3-11.

Rysin, L. L. (2012). *Forests of Moscow Region.* KMK. Moscow.

Rysin, L. P. and Savelyeva, L. I. (1994). Dynamics of pine forests on the terraces of the Moscow river. *Bulletin of Moscow Society of Naturalists. Biological series.* 99(6): 92-99.

Rysin, L. P. and Savelieva, L. I. (2002). *Spruce forests of Russia.* Nauka. Moscow.

Rysin, L. P. and Savelieva, L. I. (2008). *Pine forests of Russia.* KMK, Moscow.

Samofal, S. L. (1925). Climatic races of Scots pine (*Pinus sylvestris* L.), their importance in the organization of seed farming in the USSR. *Works on forestry.* Novaya Derevnya. Moscow. LXV (1): 5-50.

Sannikov S. N., Semerikov, L. V., Petrova, I. V. and Filippova T. V. (1997). Genetic differentiation of Scots pine populations in the Carpathians and in the Russian Plain. *Russian Journal of Ecology.* 3: 163-167.

Sannikov, S. N. and Petrova, I. V. (2003). *Differentiation of Scots pine populations.* Publishing House of the Ural Branch of the Russian Academy of Sciences. Yekaterinburg.

Sannikov, S. N. and Petrova, I. V. (2007). Phenogeography of woody plant populations: problems, methods and some results. *Conifers of the boreal zone.* XXIV (2-3): 288-294.

Sautkina, M. Yu. (2017). *Influence of associative biopreparations on the fertility of ordinary chernozem and the yield of winter triticale under the conditions of the South-East of CchZ.* Publ. VNIISS named after A.L. Mazlumov. Ramon.

Sautkina, M. Yu., Kuznetsova, N. F. and Tunyakin, V. D. (2018). Modern state of forest shelter belts with predominance of pedunculate oak (*Quercus robur* L.) in Kamennaya Steppe. *Forestry information.* 1: 78-89.

Semenov, M. A. and Kharchenko, N. N. (2017). *Mechanisms of formation of the ecosystem biological diversity at the artificial reforestation: a monograph.* Publ. VGLTU im. G.F. Morozova. Voronezh.

Semenov, M. A., Semenova, O. V., Sinitsyn, A. A. and Abramova, I. N. (2017). Modern mechanisms for achieving stability of forest biogeocenoses in the changing climate. *Modern forest science: problems and prospects.* Publ. "Istoki." Voronezh: 273-276.

Semeriakov, L. F. (1986). *Population structure of woody plants (by the example of the Pinacea family in the Urals).* Nauka. Moscow.

Schwartz, S. S. (1980). *Ecological regularities of evolution.* Nauka. Moscow.

Shkalikov, V. A. (2004). *Landscapes of the South-West of the non-Chernozem zone and their rational use (on the example of the Smolensk Region).* Publ. "Universum." Smolensk.

Sinitsin, E. M. (1982). *Indigenous and derivatives types of pine forests of the Usmansky and Khrenovskoy Bors.* Publ. VSU. Voronezh.

Smirnova, O. V. and Bobrovsky, M. V. (2000). Impact of a derivative farm on the composition and structure of forest cover. *Evaluation and conservation of forest cover biodiversity in reserves in European Russia.* Nauchnyymir. Moscow: 22-27.

Smirnova, O. V., Turubanova, M. V., Bobrovsky, M. V., Korotkov, V. M. and Khanina, L. G. (2001). Reconstruction of the history of forest cover in Eastern Europe and the problem of maintaining biological diversity. *Biology Bulletin Reviews.* 121 (2): 144-159.

Svintsova, V. S., Kuznetsova, N. F. and Pardaeva E. Yu. (2014). Influence of drought on the generative sphere and pollen viability of Scots pine. *Russian Journal of Forest Science.* 3: 49-57.

Tambov forest. (2006). Ed. N. I. Ponomarev, V. K. Shirnin. LLC "Publishing house Julis." Tambov.

Terekhin, E. A. (2016). Geoinformational mapping of changes in forests based on satellite images (using the example of Belgorod Region). *Geography and Natural Resources*. 4: 174-181.

Tsvetkov, M. A. (1957). *Changes in the forest cover of European Russia from the end of the 17th century to 1914.* Publishing House of the USSR Academy of Sciences. Moscow.

Ushatin, I. P., Mamonov, V. I. and Sidorov, V. I. (2000). To the ecological and silvicultural assessment of the Central Black Earth Region Bors. *Integration of fundamental science*. VGLTA. Voronezh. 337-341.

Ushatin, I. P. and Mamonov, D. N. (2012). The dynamics of forest regeneration processes on burnout areas in the Central Forest-Steppe. *Forestry engineering journal*. 3: 59-69.

Vidyakin, A. I. (2007). Phenetics, population structure and preservation of the genetic fund of Scots pine (*Pinus sylvestris L.*). *Conifers of the boreal zone*, XXIV (2-3): 159-166.

Vidyakin, A. I. (2014). Use of results of phenogeographical researches in practice of forestry of Russia. *Siberian Journal of Forest Science*. 4: 29-34.

Virsky, A. A. (1925). Essay of physical geography of Kursk Region. *Kursk region: collection on nature, history, culture and economy of Kursk province*. Issue 1. Publ. Kursk province of the RCP. Kursk.

Vysotsky, K. K. (1962.). *Regularities of the structure of the mixed forest stands*. Goslesbumizdat. Moscow.

Zakharov, Yu. G. (2010). Variability in the trends of linear growth from natural regeneration of pine in the conditions of the Tver Region. *Forestry bulletin*. 3: 94-96.

Zelenin, N. (2000). *Basic provisions of the organization and forest management in the territory of the Central Chernozem Areas (CCA) of Russia (The Belgorod, Voronezh, Kursk, Lipetsk, Oryol and Tambov Regions)*. FGUG "Voronezhlesproyekt." Voronezh.

Zhigunov, A. V. (2008). Priority directions of forest seed breeding and plantation forest growing in the North-West of Russia. *Forestry information*. 3-4:11-15.

Zubova, A. I. (1990). *Natural resources of the Yaroslavl Region, their protection and rational use*. 2nd ed. Upper Volga book Publ. Yaroslavl.

In: Pinus
Editor: Sylvester Stephens

ISBN: 978-1-53616-429-9
© 2019 Nova Science Publishers, Inc.

Chapter 2

PROPAGATION OF AN ENDANGERED SPECIES PINUS ARMANDII VAR. AMAMIANA VIA SOMATIC EMBRYOGENESIS

Tsuyoshi E. Maruyama[1],* and Yoshihisa Hosoi[2]
[1]Department of Research Planning and Coordination,
Forestry and Forest Products Research Institute, FFPRI, Japan
[2]Department of Forest Molecular Genetics and Biotechnology,
Forestry and Forest Products Research Institute, FFPRI, Japan

ABSTRACT

Somatic embryogenesis in Yakutanegoyou, *Pinus armandii* Franch. var. *amamiana* (Koidz.) Hatusima, an endemic and endangered species in Japan, was initiated from megagametophytes containing zygotic embryos on a medium supplemented with 2,4-dichlorophenoxyacetic acid and 6-benzylaminopurine. Embryogenic cultures were maintained and proliferated by subcultures at 2- to 3-week intervals on the same fresh medium. The maturation of somatic embryos occurred on media

* Corresponding Author's E-mail: tsumaruy@ffpri.affrc.go.jp.

containing maltose, polyethylene glycol, abscisic acid, and activated charcoal. High frequencies of germination were obtained after the post-maturation treatment of somatic embryos under conditions of high relative humidity, and after the maturation of somatic embryo on medium with a high concentration of gellan gum.

Keywords: clonal propagation, endangered pine, gellan gum, megagametophyte, *Pinus armandii*, polyethylene glycol, somatic embryogenesis, tissue culture

INTRODUCTION

Yakutanegoyou (*Pinus armandii* Franch. var. *amamiana* (Koidz.) Hatusima) is an endangered five-needle pine species endemic to Yakushima and Tanegashima Islands in the southwestern region of Japan (Environment Agency of Japan, 2000). Due to the over-harvesting of this species during the 16th to the early 20th centuries for fishing canoes and house construction, the number of trees on both islands was radically reduced. By 1994, the estimated numbers of living trees in natural stands on Tanegashima and Yakushima were 100 and 1000-1500, respectively (Yamamoto and Akashi, 1994). In recent years, Yakutanegoyou populations have further decreased as a result of pine wilt disease (Figure 1), caused by the pinewood nematode, *Bursaphelenchus xylophilus* (Steiner et Buhrer) Nickle, which is transmitted in Japan by two cerambycid beetles, namely, *Monochamus alternates* Hope (Mamiya and Enda, 1972) and *M. saltuarius* (Gebler) (Sato et al., 1987).

Pine wilt disease has been a most critical factor in the mass mortality not only of Yakutanegoyou populations but also in other important Japanese pine forests (*P. densiflora* Sieb. et Zucc., *P. thunberghii* Parl., *P. luchuensis* Mayr.). The nematode is inferred to be native to North America; since its introduction into Japan at the beginning of the 20th century, the pinewood nematode has spread to Korea, Taiwan, and China and has devastated pine forests in East Asia (Togashi and Shigesada, 2006). The nematode was also found in Portugal in 1999 (Mota et al.,

1999) and has rapidly spread to Spain and other European countries causing damage in *Pinus pinaster* forests (Nunes da Silva et al., 2015). In addition to pine wilt disease, inbreeding depression and poor reproductive characteristics have been also suggested as important causal factors of recent decline in the natural population, which denotes a high possibility of extinction in the near future (Kanetani et al., 2004). Therefore, it is essential to develop an efficient and stable plant regeneration system for large-scale propagation of resistant clones in long-term breeding programs. Somatic embryogenesis is the most attractive plant regeneration system for large-scale clonal propagation of selected trees and for *ex situ* conservation of genetic resources by cryopreservation techniques. In this chapter, we describe the initiation of embryogenic cultures from seed explants, production of somatic embryos, and regeneration of plants.

Figure 1. Mortality of Yakutanegoyou (*Pinus armandii* var. *amamiana*) trees caused by the pine wilt disease.

EMBRYOGENIC CULTURE INITIATION AND PROLIFERATION

Open-pollinated cones were collected from mother trees originated from the Kagoshima Prefectural Government Forestry Research Center (Aira, Kagoshima, Japan), from the Iso Garden (Kagoshima, Kagoshima, Japan), and from a natural population in Yakushima Island (Segire,

Kagoshima, Japan). Collected cones were disinfected by 5 min immersion in 99.5% ethanol, and then were dried in a laminar-flow cabinet before dissection. Excised seeds (Figure 2) were disinfected with 2.5% (w/v available chlorine) sodium hypochlorite solution for 30 min, and then rinsed 5 times with sterile distilled water. After the seed coats had been removed, the megagametophytes (containing zygotic embryos at the precotyledonary stage) were used as explants for the initiation of embryogenic cultures.

The explants were cultured in multiwell plates containing initiation medium (1/2-EM medium) (Maruyama et al., 2000), modified as follows: salts, vitamins, and myo-inositol were reduced to half the standard concentration; the concentration of KCl was reduced to 40 mg l^{-1}; and 500 mg l^{-1} casein hydrolysate was added, as well as 1 g l^{-1} L-glutamine, 10 g l^{-1} sucrose, 10 μM 2,4-dichlorophenoxyacetic acid (2,4-D), and 5 μM 6-benzylaminopurine (BA). The medium was solidified with 3 g l^{-1} gellan gum (Gelrite®; Wako Pure Chemical, Osaka, Japan). The pH of the medium was adjusted to 5.8 prior to sterilization. Cultures were kept in darkness at approximately 25°C. The same medium without plant growth regulators (PGRs) but containing 2 g l^{-1} activated charcoal (AC) (Wako Pure Chemical, Osaka, Japan) was also tested for the induction of embryogenic cells. The presence or absence of embryogenic cells was determined weekly under an inverted microscope up to 12 weeks.

Figure 2. Seeds of Yakutanegoyou (*Pinus armandii* var. *amamiana*).

Table 1. Embryogenic response of 9 open-pollinated seed families from megagametophytes containing zygotic embryos of Yakutanegoyou (*Pinus armandii* var. *amamiana*)

Mother tree	Induction medium tested	Number of explant tested	Explant with embryogenic tissue	Induction frequency %
ISO	-PGR	39	0	0
	+PGR	108	0	0
HR75	-PGR	12	0	0
	+PGR	14	0	0
HR17	-PGR	18	0	0
	+PGR	18	1	5.6
T10	-PGR	5	1	20
	+PGR	3	0	0
H11TM	+PGR	7	0	0
T10TM	+PGR	9	0	0
I01	-PGR	24	0	0
	+PGR	48	1	2.1
I25	-PGR	51	1	2
	+PGR	75	0	0
HR10	-PGR	15	0	0
	+PGR	15	3	20
Total	-PGR	164	2	1.2
	+PGR	297	5	1.7
	Total	461	7	1.5

-PGR or +PGR: induction medium without or with plant growth regulators.

Figure 3. Induced embryogenic tissue from seed of Yakutanegoyou (*Pinus armandi* var. *amamiana*).

To promote proliferation, embryogenic tissues (ETs) were transferred onto fresh initiation medium supplemented with 30 g l^{-1} sucrose, 3 μM 2,4-D, and 1 μM BA, and without casein hydrolysate. Once embryogenic cultures increased in mass, 5-10 pieces of tissue per plate (90 x 20 mm) were maintained and proliferated by subculturing at 2- to 3-week intervals and incubated under the same conditions as for initiation.

The extrusion of ETs from the micropylar end of explants occurred mostly after 4–6 weeks of culture (Figure 3). As shown in Table 1, an average of 1.5% (7/461) of megagametophytes tested had extruded the zygotic embryos and then proliferated into embryonal masses that became evident after approximately 8 weeks of culture. Five explants from 297 megagametophytes (1.7%) and 2 from 164 (1.2%) with proliferating ETs were noted on medium with and without PGRs, respectively.

ETs were removed from the explants and transferred to maintenance-proliferation medium. A total of seven embryogenic cultures, established from 461 explants (1.5%), have been maintained over 2 years by subculturing at 2- to 3-week intervals on medium supplemented with 2,4-D and BA. ETs proliferated readily and retained their original translucent and mucilaginous appearance (Figure 4). The fresh weight of tissue on maintenance-proliferation medium increased 8- to 15-fold after a 2- to 3-week culture period. The EM medium at half strength supplemented with 3 μM 2,4-D and 1 μM BA supported the growth of the embryogenic cell lines tested.

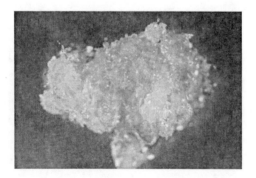

Figure 4. Embryogenic tissue proliferation in Yakutanegoyou (*Pinus armandii* var. *amamiana*).

MATURATION OF SOMATIC EMBRYOS

ETs, 14 days after subculture, were used for maturation experiments. About 500 mg fresh weight suspended in about 3 ml of liquid medium (medium of the same composition used for the maintenance and proliferation, but without PGRs) was poured over each 90 x 20 mm plates containing 30-40 ml of semi-solid maturation medium. Maturation medium contained salts and vitamins from the original EM medium, 50 g l^{-1} maltose, 100 μM abscisic acid (ABA), 0-150 g l^{-1} polyethylene glycol 4000 (Av. Mol. Wt.: 3000; Wako Pure Chemical, Osaka, Japan) (PEG), 0-2 g l^{-1} AC, 3 g l^{-1} gellan gum, and EMM amino acids (Smith, 1996) (g l^{-1}: glutamine 7.3, asparagine 2.1, arginine 0.7, citrulline 0.079, ornithine 0.076, lysine 0.055, alanine 0.04 and proline 0.035). The pH of the medium was adjusted to 5.8 prior to sterilization. Petri dishes were sealed with Novix-II film (Iwaki Glass, Chiba, Japan) and kept in darkness at approximately 25°C for 8–12 weeks.

Table 2. Effect of PEG and AC on cotyledonary somatic embryo formation in Yakutanegoyou (*Pinus armandii* var. *amamiana*)

Media	PEG (g l^{-1})	AC (g l^{-1})	Somatic embryos per plate (SE)
P0	0	0	19 (6) a
P0AC	0	2	24 (7) a
P50	50	0	22 (5) a
P50AC	50	2	41 (13) ab
P75	75	0	33 (9) a
P75AC	75	2	79 (18) bc
P100	100	0	40 (9) ab
P100AC	100	2	101 (31) c
P150	150	0	18 (7) a
P150AC	150	2	45 (17) ab

SE, standard errors of means from five replications for each treatment.

Means followed by same letter are not significantly different at a level P < 0.05.

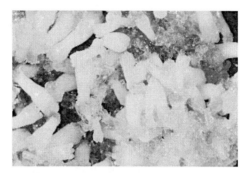

Figure 5. Cotyledonary somatic embryo formation in Yakutanegoyou
(*Pinus armandii* var. *amamiana*).

**Table 3. Somatic embryo production from embryogenic cell lines
of Yakutanegoyou (*Pinus armandii* var. *amamiana*)**

Cell line	Average number of somatic embryos per plate (SE)
HR17-A	101 (31)
T10-1	0
I01-1	149 (42)
I25-5	24 (8)
HR10-A	169 (31)
HR10-B	20 (4)
HR10-C	8 (4)

SE, standard errors of means from five replications for each cell line.

Somatic embryo development in six of the seven lines tested was
promoted on maturation media. Cotyledonary somatic embryos were
observed about 4 weeks after transfer of ETs, and were distinct at 6-8
weeks of culture (Figure 5). The addition of PEG to the medium stimulated
the embryo maturation, and this effect was even more enhanced in the
presence of AC (Table 2). The number of mature embryos increased with
increased PEG concentration up to 100 g l^{-1}, but decreased at 150 g l^{-1}. In
contrast, on PEG-free medium, the embryogenic cell proliferation was
evident and only a few early somatic embryos developed into cotyledonary
stage.

Although 75 and 100 g l^{-1} of PEG did not result in a statistical
difference in terms of the somatic embryo yield per plate, the highest

embryo maturation frequency was obtained on medium supplemented with 100 g l⁻¹ PEG, 2 g l⁻¹ AC, and containing 100 μM ABA and 50 g l⁻¹ maltose as a carbohydrate source (P100AC). Therefore, for further experiments on somatic embryo maturation with different cell lines, P100AC medium was used.

The results on embryo maturation for 7 different cell lines are shown in Table 3. As observed in our earlier experiments, in most cell lines the development of somatic embryos to the cotyledonary stage was evident after 6 wk of culture. Somatic embryos were produced in six of seven cell lines tested and the ability to produce mature somatic embryos was notably different among the cell lines. The average number of somatic embryos per plate ranged from 8 to 169.

SOMATIC EMBRYO GERMINATION, PLANTLET CONVERSION, AND ACCLIMATIZATION OF SOMATIC PLANTS

Cotyledonary somatic embryos collected from maturation medium after 8 weeks of culture were transferred directly to germination medium or subjected to partial desiccation prior to germination. Partial desiccation of somatic embryos was carried out as described by Maruyama and Hosoi (2012). Briefly, somatic embryos were placed on filter paper sheets equipped in 2 (central) wells of a 6-well multiplate (Iwaki, Asahi Glass Co., Ltd, Tokyo, Japan) in which the remaining 4 (side) wells were filled with 5−6 ml of sterile water, sealed tightly with Parafilm, and kept in darkness at 25°C for 3 weeks. Under these conditions, the generated relative humidity inside the plate was approximately 98%, which was registered with a thermo-hygrometer recorder (RS-10, ESPEC MIC Corp. Aichi, Japan). The germination medium was of the same composition as for the maintenance and proliferation, but without PGRs and glutamine; and supplemented with 20 g l⁻¹ sucrose, 2 g l⁻¹ AC, and 10 g l⁻¹ agar (Wako Pure Chemical, Osaka, Japan). Cultures were kept at approximately 25°C under a photon flux density of about 65 µmol m^{-2} s^{-1} provided by cool, white fluorescent lamps (100V, 40W; Toshiba, Tokyo, Japan) for 16 h. The numbers of somatic embryos germinated (root emergence) and converted into plantlets (emergence of both root and epicotyl) were recorded after 6 and 12 weeks, respectively.

Regenerated plantlets were transferred into 300-ml flasks containing 100 ml of fresh medium (1/2-EM medium without PGRs and containing 30 g l^{-1} sucrose, 5 g l^{-1} AC and 10 g l^{-1} agar) or into Magenta® vessels (Sigma, St. Louis, USA) containing Florialite® (Nisshinbo Industries, Tokyo, Japan) irrigated with a 0.1% (v/v) Hyponex® 6-10-5 plant food solution (Hyponex Japan Co., Ltd., Osaka, Japan), and kept under the same conditions as described above. Subsequently, developed plants were transplanted into plastic pots filled with vermiculite or Kanuma soil and acclimatized in a growth cabinet as described by Maruyama et al., (2002).

Table 4. Germination and conversion frequencies from somatic embryos of Yakutanegoyou (*Pinus armandii* var. *amamiana*) after maturation with PEG or a high concentration of gellan gum

Cell line	Germination frequency (%)			Conversion frequency (%)		
	PEG[1]	PEG + Desiccation[2]	Gellan gum[3]	PEG[1]	PEG + Desiccation[2]	Gellan gum[3]
HR-17A	46	85	83	31	80	82
HR-10A	40	89	91	35	88	90
HR-10B	32	73	77	28	70	75
HR-10C	36	80	82	34	80	82
I01-1	51	91	97	40	90	95
I25-5	16	55	51	12	54	50
Total	41	83	81	33	80	80

[1] Somatic embryos generated on maturation medium supplemented with PEG.

[2] Somatic embryos generated on maturation medium supplemented with PEG were partial desiccated at high relative humidity.

[3] Somatic embryos generated on maturation medium supplemented without PEG and high concentration of gellan gum.

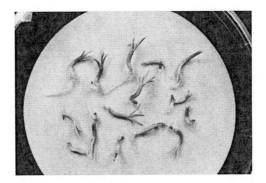

Figure 6. Somatic embryo germination in Yakutanegoyou
(*Pinus armandii* var. *amamiana*).

Figure 7. Yakutanegoyou (*Pinus armandii* var. *amamiana*) somatic plants
growing *in vitro.*

Figure 8. Acclimatized somatic plants of Yakutanegoyou
(*Pinus armandii* var. *amamiana*).

Figure 9. Yakutanegoyou (*Pinus armandii* var. *amamiana*) somatic plant before transfer to the field.

Figure 10. Yakutanegoyou (*Pinus armandii* var. *amamiana*) somatic plants growing in the field.

As shown in Table 4, when somatic embryos matured with PEG were placed directly on the germination medium, the root emergence of embryos and the subsequent plant conversion occurred at a low frequency (an average of 41% and 33%, respectively). In contrast, when the somatic embryos were matured on medium containing a high concentration of gellan gum without PEG, the germination frequency recorded was 81%, and then 80% of somatic embryos developed into plantlets. The start of germination (root emergence) was observed 1–2 weeks after transfer into the germination medium (Figure 6), and the embryos subsequently converted into plantlets (emergence of both root and epicotyl) after 4–8 weeks of culture.

Similarly, partial desiccation of somatic embryos resulted not only in a marked increment in the germination frequencies but also in subsequent improvement of plant conversion rates in all the cell lines tested (Table 4).

Regenerated somatic plants were cultured *in vitro* for 12 to 20 weeks before acclimatization *ex vitro* (Figure 7). Subsequently, the acclimatized plants (Figure 8) were transferred to a greenhouse and grown for several months (Figure 9) before transplanting to the field. The growth of somatic plants is currently being monitored in the field (Figure 10).

DISCUSSION AND CONCLUSION

Somatic embryogenesis of Yakutanegoyou was initiated in five of the nine seed families tested, with a total average of 1.5% (Table 1). This rate was low but consistent with reported results for *P. banksiana* (0.4%, Park et al., 1999), *P. thunbergii* (2%) and *P. densiflora* (1%) (Ishii et al., 2001), *P. rigida* × *P. taeda* (up to 1.1%, Kim and Moon 2007), and *P. lambertiana* (1%–3%, Gupta 1995). In contrast, higher initiation rates were reported for *P. taeda* (up to 79%, Gupta 2014), *P. strobus* (54%, Finer et al., 1989), *P. sylvestris* (up to 30%, Aronen et al., 2009) and *P. pinaster* (up to 75%, Park et al., 2006). On the other hand, Kim et al., (1999) reported that from 294 lines initiated in *Larix leptolepis*, only one embryogenic cell line could be proliferated. In the present study, although the initiation frequencies for Yakutanegoyou were relatively low, all of initiated lines continued to proliferate even after 2 years of culture, resulting in stable embryogenic lines. These results suggested that the capture of stable cell lines was the most appropriate criterion by which to compare the ability of somatic embryogenesis initiation among species and families (Maruyama et al., 2007). Initiation of ETs (Table 1) was possible also in the absence of exogenous growth regulators as reported for *Pinus radiata* (Smith, 1996), *P. sylvestris*, *P. pinaster* (Lelu et al., 1999), *P. thunberghii* and *P. densiflora* (Ishii et al., 2001). These results suggest that when explants are cultured at the appropriate developmental stage, the megagametophyte tissue may supply the PGRs necessary for *in vitro* somatic embryogenesis initiation from zygotic embryos without a requirement for additional supplement of exogenous auxin and cytokinin. However, exogenous PGRs were found essential for the maintenance and proliferation of ETs. This result was consistent with our previous results for other Japanese conifers (Maruyama et al., 2000, 2002, 2005).

Development and maturation patterns of Yakutanegoyou were similar to those described for other pines (Becwar et al., 1990; Smith, 1996; Lelu et al., 1999). Embryo maturation efficiency was promoted by culturing embryogenic tissue on media supplemented with maltose, PEG, ABA, and AC. We determined in preliminary experiments (data not shown) that the presence of ABA was essential for the production of somatic embryos in the presence of PEG. ABA-free medium failed to stimulate somatic embryo maturation, because only a few cotyledonary embryos developed. This result is consistent with reports on *P. strobus* (Klimaszewska and Smith, 1997), *P. pinaster* and *P. sylvestris* (Lelu et al., 1999). In those species, a higher concentration of ABA resulted in a higher yield of cotyledonary somatic embryos. Absence or lower concentrations of ABA resulted in the production of fewer embryos, most of which were abnormal. Similarly, somatic embryos of hybrid larch (*Larix* × *leptoeuropaea*) developed normally on a medium supplemented with ABA, but abnormally on a medium with no ABA (Gutmann et al., 1996). Most of the studies of somatic embryogenesis in conifers have concluded that ABA was very important in embryo development, and that the quantity and quality of embryos was inferior in the absence of ABA (Attree and Fowke, 1993; Dunstan et al., 1998). Several authors have suggested that the positive effect of ABA on somatic embryo development in conifers is closely related to the inhibition of cleavage polyembryony and to the accumulation of storage lipids, proteins and carbohydrates during maturation (Roberts et al., 1990; Gupta et al., 1991).

PEG in combination with ABA has been used for somatic embryo maturation in several conifer species. The stimulating effect of PEG on embryo maturation may be related to a water stress induction similar to that generated by desiccation, due to its non-plasmolysing characteristics and to an increase in the accumulation of storage reserves and triglycerides (Attree and Fowke, 1993). The beneficial effect of AC on somatic embryo development was also corroborated in this study (Table 2). AC has been used, mostly in combination with ABA and PEG, in somatic embryogenesis of gymnosperms (Jain et al., 1995; Maruyama et al., 2002; Jain and Gupta, 2005). AC is widely used in tissue culture media, where it is believed to function as an adsorbent for toxic metabolic products and residual hormones (Pullman and Gupta, 1991).

Although the ability of Yakutanegoyou to produce mature somatic embryos varied according to the genotype, as shown in Table 3, the somatic embryo production in 50% of genotypes that responded (3 of 6 lines) was over 100 cotyledonary embryos per plate (101 to 169 on average). Variation in the potential for mature embryo production and subsequently for plant regeneration is commonly reported among genotypes and families in conifers (Lelu et al., 1999; Igasaki et al., 2003; Maruyama et al., 2005). Somatic embryo production in other pine species averaged up to 110 embryos per Petri dish in *P. radiata* (Smith, 1996), up to 295 and 262 embryos per gram f. w. in *P. strobus* (Klimaszewska and Smith, 1997) and *P. sylvestris* (Lelu et al., 1999), respectively; and up to 231 embryos per ml of settled cells in *P. taeda* (Pullman et al., 2003).

The maturation frequency and the quality of somatic embryos produced are the most important criteria for the optimization of a somatic embryogenesis protocol. The quality of a somatic embryo is related to its morphology, biochemical similarity to zygotic embryo, and the ability to produce normal plants. High-quality somatic embryos have a zygotic embryo-like morphology and generally have a radial symmetry. In experiments with Yakutanegoyou, we obtained somatic embryos resembling zygotic embryos, mostly when ETs were cultured on P100AC maturation medium or on medium without PEG but containing a high concentration of gellan gum. In contrast, somatic embryos with stunted hypocotyls are generally considered low-quality. These embryos, referred to as shooty embryos (Lelu et al., 1999), had abnormal morphology, which might have led to a low efficiency of embryo germination. In order to improve plant conversion protocols, the desiccation of somatic embryos after PEG-mediated maturation has been recommended for conifer species. In Yakutanegoyou, partial desiccation of cotyledonary embryos considerably increased the germination rates and also improved the subsequent plant conversion frequencies (Table 4). For most species, desiccation presumably acts to terminate developmental processes and to initiate those metabolic processes necessary to prepare the seeds for germination and growth (Kermode and Bewley 1985). Although the specific changes were not explored in this study, the improved performance of somatic embryos after desiccation treatment can be attributed to a change in endogenous hormone levels and accumulation of storage reserves (Ackerson 1984, Dronne et al., 1997, Kong and Yeung 1992, Find 1997, Stasolla et al., 2001, Klimaszewska et al., 2004). The beneficial effect of desiccation treatment after maturation with PEG, improving the germination frequencies and decreasing the time required for germination, was also reported for somatic embryos of interior spruce (Roberts et al., 1990), sitka spruce (Roberts et al., 1991), white spruce (Kong and Yeung 1992, Kong and Yeung 1995, Attree et al., 1995), hybrid larch (Lelu et al., 1995, Dronne et al., 1997), patula pine (Jones and van Staden 2001), and Ryukyu pine (Hosoi and Maruyama 2012).

On the other hand, somatic embryo maturation methods, involving reduction in water availability to the cultured cells by increasing the medium gel strength (with a high concentration of gellan gum) to produce mature somatic embryos with low water content, have been reported in several improved protocols for pine species (Maruyama and Hosoi, 2019). Restricting water availability has resulted in high germination rates and subsequent high plant conversion frequencies in *P. radiata* (Smith 1996),

P. strobus (Klimaszewska and Smith 1997, Klimaszewska et al., 2001), *P. sylvestris* (Lelu et al., 1999), *P. monticola* (Percy et al., 2000), *P. pinaster* (Lelu et al., 1999, Lelu-Walter et al., 2006), and *P. halepensis* (Montalban et al., 2013). In Yakutanegoyou, although the somatic embryo production on PEG-free medium containing a high concentration of gellan gum (10 g l^{-1}) was lower than that achieved on PEG-supplemented media (data not shown), the obtained cotyledonary somatic embryos showed a high germination frequency without any post-maturation treatments (Table 4).

In conclusion, although the described regeneration system represents a promising perspective for efficient propagation of this species, further studies including protocol improvement for somatic embryogenesis initiation and monitoring of genetic stability of somatic embryos, are required to establish a practical and effective regeneration method for the large-scale production of high quality somatic plants resistant to pine wilt disease.

ACNOWLEDGMENTS

We express our gratitude to the Kagoshima Prefectural Government Forestry Research Center and to the Iso Garden for the generous supply of seeds.

REFERENCES

Ackerson, R. C. (1984). Abscisic acid and precocious germination in soybeans. *J Exp Bot* 35:414-421

Aronen, T., Pehkonen, T. & Ryynänen, L. (2009). Enhancement of somatic embryogenesis from immature zygotic embryos of *Pinus sylvestris. Scand J For Res* 24:372-383.

Attree, S. M. & Fowke, L. C. (1993). Somatic embryogenesis and synthetic seeds of conifers. *Plant Cell Tiss Organ Cult* 35:1–35.

Attree, S. M., Pomeroy, M. K. & Fowke, L. C. (1995). Development of white spruce (*Picea glauca* (Moench.) Voss) somatic embryos during culture with abscisic acid and osmoticum, and their tolerance to drying and frozen storage. *J Exp Bot* 46: 433-439.

Becwar, M. R., Nagmani, R. &Wann, S. R. (1990). Initiation of embryogenic cultures and somatic embryo development in loblolly pine (*Pinus taeda*). *Can J For Res* 20:810–817.

Dronne, S., Label, P. & Lelu, MA. (1997). Desiccation decreases abscisic acid content in hybrid larch (*Larix x leptoeuropaea*) somatic embryos. *Physiol Plant* 99:433-438.

Dunstan, D. I., Bekkaoui, F., Pilon, M., Fowke, L. C. & Abrams, S. R. (1998). Effects of abscisic acid and analogues on the maturation of white spruce (*Picea glauca*) somatic embryos. *Plant Sci* 58:77–84.

Environment Agency of Japan (2000). *Threatened wildlife of Japan-Red Data Book*, 2nd ed. vol. 8, Vascular Plants. Japan Wildlife Research Center, Tokyo (in Japanese).

Find, J. I. (1997). Changes in endogenous ABA levels in developing somatic embryos of Norway spruce (*Picea abies* [L.] Karts.) in relation to maturation medium, desiccation and germination. *Plant Sci* 128:75-83.

Finer, J. J., Kriebel, H. B. & Becwar, M. R. (1989). Initiation of embryogenic callus and suspension cultures of eastern white pine (*Pinus strobus* L.). *Plant Cell Rep* 8:203–206.

Gupta, P. K. (1995). Somatic embryogenesis in sugar (*Pinus lambertiana* Dougl.). In: Jain, S. M., Gupta, P. K. & Newton, R. J. (eds) *Somatic embryogenesis in woody plants Vol. 3* (pp. 197-205). Kluwer Academic Publishers, Dordrecht.

Gupta, P. K. (2014). *Methods of initiating plant somatic embryos*. WIPO/PCT WO 2014/100102 A1.

Gupta, P. K., Timmis, R., Pullman, G., Yancey, M., Kreitinger, M., Carlson, W. & Carpenter, C. (1991). Development of an Embryogenic System for Automated Propagation of Forest Trees. In: Vasil, I. K., ed. *Cell culture and somatic cell genetics of plants, vol. 8* (pp. 75-93). Academic Press Inc., San Diego.

Gutmann, M., von Aderkas, P., Label, P. & Lelu, M. A. (1996). Effects of abscisic acid on somatic embryo maturation of hybrid larch. *J Exp Bot* 35:1905–1917.

Hosoi, Y. & Maruyama, T. E. (2012). Plant regeneration from embryogenic tissue of *Pinus luchuensis* Mayr, an endemic species in Ryukyu Island, Japan. *Plant Biotech* 29:401-406.

Igasaki, T., Sato, T., Akashi, N., Mohri, T., Maruyama, E., Kinoshita, I., Walter, C. & Shinohara, K. (2003). Somatic embryogenesis and plant regeneration from immature zygotic embryos of *Cryptomeria japonica* D. Don. *Plant Cell Rep* 22:239–243.

Ishii, K., Maruyama, E. & Hosoi, Y. (2001). Somatic embryogenesis of Japanese conifers. In: Morohoshi, N. & Komamine, A. eds. *Molecular breeding of woody plants* (pp. 297–304). Elsevier Science, Amsterdam.

Jain, S. M. & Gupta, P. K. (2005). *Protocol for somatic embryogenesis in woody plants.* Springer, Dordrecht.

Jain, S. M., Gupta, P. K. & Newton, R. J. (1995). *Somatic embryogenesis in woody plants, vol. 3,* Gymnosperms. Kluwer Academic Publishers, Dordrecht.

Jones, N. B. & van Staden, J. (2001). Improved somatic embryo production from embryogenic tissue of *Pinus patula. In Vitro Cell Dev Biol-Plant* 37:543-549.

Kanetani, S., Kawahara, T., Kanazashi, A. & Yoshimaru, H. (2004). Diversity and conservation of an endangered five-needle pine species, *Pinus armandii* Franch. var *amamiana* (Koidz.) Hatusima. In: *Proceedings of breeding and genetic resources of five-needle pines: Growth, adaptability and pest resistance* (pp. 188-191). *USDA Forest Service Proceedings RMRS-P-32.*

Kermode, A. R. & Bewley, D. (1985). The role of maturation drying in the transition from seed development to germination. I. Acquisition of desiccation-tolerance and germinability during development of *Ricinus communis* L. seeds. *J Exp Bot* 12:1906-1915.

Kim, Y. W. & Moon, H. K. (2007). Regeneration of plant by somatic embryogenesis in *Pinus rigida* × *P. taeda*. *In Vitro Cell Dev Biol-Plant* 43: 335-342.

Kim, Y. W., Youn, Y., Noh, E. R. & Kim, J. C. (1999). Somatic embryogenesis and plant regeneration from immature zygotic embryos of Japanese larch (*Larix leptolepis*). *Plant Cell Tiss Organ Cult* 55:95–101.

Klimaszewska, K. & Smith, D. R. (1997). Maturation of somatic embryos of *Pinus strobus* is promoted by a high concentration of gellan gum. *Physiol Plant* 100:949–957.

Klimaszewska, K., Morency, F., Jones-Overton, C. & Cooke, J. (2004). Accumulation pattern and identification of seed storage proteins in zygotic embryos of *Pinus strobus* and in somatic embryos from different maturation treatments. *Physiol Plant* 121: 682-690.

Klimaszewska, K., Park, J. S., Overton, C., Maceacheron, I. & Bonga, J. M. (2001) Optimized somatic embryogenesis in *Pinus strobus* L. *In Vitro Cell Dev Biol. Plant* 37:392–399.

Kong, L. & Yeung, E. C. (1995). Effects of silver nitrate and polyethylene glycol on white spruce (*Picea glauca*) somatic embryo development: enhancing cotyledonary embryo formation and endogenous ABA content. *Physiol Plant* 93:298-304.

Kong, L. &, Yeung, E. C. (1992). Development of white spruce somatic embryos: II. Continual shoot meristem development during germination. *In Vitro Cell Dev Biol-Plant* 28P:125-131.

Lelu, M. A., Bastien, C., Drugeault, A., Gouez, M. L. & Klimaszewska, K. (1999). Somatic embryogenesis and plantlet development in *Pinus sylvestris* and *Pinus pinaster* on medium with and without growth regulators. *Physiol Plant* 105:719–728.

Lelu, M-A., Klimaszewska, K., Pflaum, G. & Bastien, C. (1995). Effect of maturation duration on desiccation tolerance in hybrid larch (*Larix x leptoeuropaea* Dengler) somatic embryos. *In Vitro Cell Dev Biol-Plant* 31: 15-20.

Lelu-Walter, M. A., Bernier-Cardou, M. & Klimaszewska, K. (2006). Simplified and improved somatic embryogenesis for clonal propagation of *Pinus pinaster*. *Plant Cell Rep* 25:767-776.

Mamiya, Y. & Enda, N. (1972). Transmission of *Bursaphelenchus lignicolus* (Nematoda: Aphelenchoididae) by *Monochamus alternus* (Coleoptera: Cerambycidae). *Nematologica* 18:159-162.

Maruyama, E., Hosoi, Y. & Ishii, K. (2002). Somatic embryogenesis in Sawara cypress (*Chamaecyparis pisifera* Sieb. et Zucc.) for stable and efficient plant regeneration, propagation and protoplast culture. *J For Res* 7:23–34.

Maruyama, E., Hosoi, Y. & Ishii, K. (2007). Somatic embryogenesis and plant regeration in Yakutanegoyou, *Pinus armandii* Franch. var. *amamiana* (Koidz.) Hatusima, an endemic and endangered species in Japan. *In Vitro Cell Dev Biol-Plant* 43:28-34.

Maruyama, E., Ishii, K. & Hosoi, Y. (2005). Efficient plant regeneration of Hinoki cypress (*Chamaecyparis obtusa* Sieb. et Zucc.) via somatic embryogenesis. *J For Res* 10:73-77.

Maruyama, E., Tanaka, T., Hosoi, Y., Ishii, K. & Morohoshi, N. (2000). Embryogenic cell culture, protoplast regeneration, cryopreservation, biolistic gene transfer and plant regeneration in Japanese cedar (*Cryptomeria japonica* D. Don). *Plant Biotechnol* 17:281–296.

Maruyama, T. E. & Hosoi, Y. (2012). Post-maturation treatments improve and synchronizes somatic embryo germination of three species of Japanese pines. *Plant Cell Tiss Organ Cult* 110:45-52.

Maruyama, T. E. & Hosoi, Y. (2019). Progress in somatic embryogenesis of Japanese pines. *Front Plant* Sci 10:31.

Montalbán, L. A., Setien-Olarra, A., Hargreaves, C. L. & Moncaleán, P. (2013). Somatic embryogenesis in *Pinus halepensis* Mill.: an important ecological species from the Mediterranean forest. *Trees* 27:1339-1351.

Mota, M. M., Braasch, H., Bravo, M. A., Penas, A. C., Burgermeister, W., Metge, K. & Sousa, E. (1999). First report of *Bursaphelenchus xylophilus* in Portugal and in Europe. *Nematology* 1:727–734.

Nunes da Silva, M., Solla, A., Sampedro, L., Zas, R. & Vasconcelos, M. (2015). Susceptibility to the pinewood nematode (PWN) of four pine species involved in potential range expansion across Europe. *Tree Physiol* 35:987-999.

Park, J. S., Bonga, J. M., Cameron, S. I., Barrett, J. D., Forbes, K., DeVerno, L. L. & Klimaszewska, K. (1999). Somatic embryogenesis in jack pine (*Pinus banksiana* Lamb). In: Jain, S. M., Gupta, P. K. & Newton, R. J., eds. *Somatic embryogenesis in woody plants*, vol. 4 (pp. 491-504). Kluwer Academic Publishers, Dordrecht.

Park, Y. S., Lelu-Walter, M. A., Harvengt, L., Trontin, J. F., MacEacheron, I., Klimaszewska, K. & Bonga, J. M. (2006). Initiation of somatic embryogenesis in *Pinus banksiana*, *P. strobus*, *P. pinaster*, and *P. sylvestris* at three laboratories in Canada and France. *Plant Cell Tiss Organ Cult* 86:87-101.

Percy, R. E., Klimaszewska, K. & Cyr, D. R. (2000). Evaluation of somatic embryogenesis for clonal propagation of western white pine. *Can J For Res* 30:1867-1876.

Pullman, G. S. & Gupta, P. K. (1991). Method for reproducing coniferous plants by somatic embryogenesis using absorbent materials in the development stage media. *U.S. Patent No. 5 034 326.*

Pullman, G. S., Johnson, S., Peter, S., Cairney, J. & Xu, N. (2003). Improving loblolly pine somatic embryo maturation: comparison of somatic and zygotic embryo morphology, germination, and gene expression. *Plant Cell Rep* 21:747-758.

Roberts, D. R., Flinn, B. S., Webb, D. T., Webster, F. B. & Sutton, B. C. S. (1990). Abscisic acid and indole-3-butyric acid regulation of maturation and accumulation of storage proteins in somatic embryos of interior spruce. *Physiol Plant* 78:355-360.

Roberts, D. R., Lazaroff, W. R & Webster, F. B. (1991). Interaction between maturation and high relative humidity treatments and their effects on germination of Sitka spruce somatic embryos. *J Plant Physiol* 138:1-6.

Sato, H., Sakuyama, T. & Kobayashi, M. (1987). Transmission of *Bursaphelenchus xylophilus* (Steiner et Buhrer) Nickle (Nematoda, Aphelenchoididae) by *Monochamus saltuarius* (Gebler) (Coleoptera,

Cerambycidae). *J Jpn For Soc* 69:492-496 (in Japanese with English summary).

Smith, D. R. (1996). Growth medium. *United States Patent # 5,565,355.*

Stasolla, C., Loukanina, N., Ashihara, H., Yeung, E. C. & Thorpe, T. E. (2001). Purine and pyrimidine metabolism during the partial drying treatment of white spruce (*Picea glauca*) somatic embryos. *Physiol Plant* 111:93-101.

Togashi, K. & Shigesada, N. (2006) Spread of the pinewood nematode vectored by the Japanese pine sawyer: modeling and analytical approaches. *Popul Ecol* 48:271–283.

Yamamoto, C. & Akashi, T. (1994). Preliminary report on distribution and conservation of a rare tree species of *Pinus armandii* Franch. var *amamiana* (Koidz.) Hatusima. *Trans Annu Meet Jpn For Soc* 105:750 (in Japanese).

In: Pinus
Editor: Sylvester Stephens

ISBN: 978-1-53616-429-9
© 2019 Nova Science Publishers, Inc.

Chapter 3

AN APPROACH ON THE USE OF PINUS (*PINUS ELLIOTTII*) BARK AS AN ALTERNATIVE IN THE REMOVAL OF TOXIC METALS FROM WATER

Affonso Celso Gonçalves Jr.[1,*], Andréia da Paz Schiller[2],
Elio Conradi Jr.[3], Jéssica Manfrin[2],
Juliano Zimmermann[4] and Daniel Schwantes[5]
[1]Industrial Chemistry (UFSM-RS-Brazil); Master in Agrochemical
(UEM-PR-Brazil); Doctor in Analytical Chemistry (UFSC-SC-Brazil);
Post Doctor in Environmental Sciences (UFG-GO-Brazil);
Post Doctor in Remediation of Environmental Compartments
(University of Santiago de Compostela – Spain);
Post Doctor in Agricultural Sciences (UEM-PR-Brazil). Researcher in
productivity by CNPq and Associated Professor
in Center of Agricultural Sciences of State University
of West Paraná (UNIOESTE-PR-Brazil)
[2]Environmental Engineer (PUCPR-PR-Brazil);
Master in Agronomy (UNIOESTE-PR-Brazil)

---

[*] Corresponding Author's E-mail: affonso133@hotmail.com.

[3]Agronomist Engineer (UNIOESTE-PR-Brazil);
Master Student in Agronomy (UNIOESTE-PR-Brazil)
[4]Agronomist Engineer (UNIOESTE-PR-Brazil)
[5]Agronomist Engineer (UNIOESTE-PR-Brazil); Master in Agronomy
(UNIOESTE-PR-Brazil); Doctor in Agronomy (UNIOESTE-PR-
Brazil); Post Doctor in Agronomy (UNIOESTE-PR-Brazil);
Assistant Professor at Departamento de Ciencias Vegetales,
Facultad de Agronomía e Ing. Forestal
(Pontificia Universidad Católica de Chile - Chile)

ABSTRACT

The intensification of human activities generates an increasing demand for products of forest origin. The *Pinus* sp. genus, for example, has been extensively explored and consequently, it has generated a large amount of wastes (barks and sawdust from wood extraction). If they are not well managed, the barks generated by pinus can be harmful to the environment, since they store large amounts of toxic metals. In this context, the destination of the barks to obtain adsorbents for remediate toxic metals from contaminated water can be an alternative. Therefore, the present study aimed to gather information about the pinus cultivation, as well the potential use of the pinus bark as adsorbent to remove Cd^{2+}, Pb^{2+} and Cr^{3+} from contaminated water. It was done a search about the pinus cultivation in the world, focusing for studies that aimed to produce adsorbents from pinus bark and the results obtained in the remediation of contaminated water. Based in the data collection in this work, it can be conclude that the modified pinus barks become an excellent alternative in the removal of toxic metals from water (Cd^{2+}, Pb^{2+} and Cr^{3+}). In addition, the use of pinus bark as an adsorbent material represents a sustainable practice (appropriate destination of the pinus bark and the decontamination of the water), which complements the final productive steps of this species. Moreover, the adsorption process is deeply influenced by chemical compounds that substantially change the characteristics of the material and increase the adsorption.

Keywords: adsorption, alternative adsorbents, water contamination, alternative treatments of water

INTRODUCTION

The world's forest resources have been submitted in great demand for forest products, such as wood to generate heat; pulp industry; paper; wood and furniture (Choudhury et al. 2014).

Some crops grow well in tropical climate conditions; however, they do not tolerate the frost that occurs in certain locations, for example, in the southern states of Brazil. Therefore, the pinus is a good choice of cultivation for colder environments (Souza et al. 2004).

The *Pinus* sp. genus is represented by a large number of native species of the Northern Hemisphere, in American, European and Asian continents and in the Northern part of Africa. Although there are shrub species, mainly in the semi-arid regions, and some with low characteristics (occurring in high altitude locations), most of them produce large erect trees. These are important sources of long fiber wood, usually obtained from natural forests as well as the intensive forestry cultivation and managed plantations (as a way to optimize production costs and maximize returns on the generation of high quality in the raw material) (Shimizu et al. 2017).

In consequence, to the high pinus production, it has generated a high amount of waste (bark and sawdust). These wastes could have adverse environmental characteristics, since when burned they can generate toxic waste, such as combustion gases and ash with high concentrations of toxic metals (Delucius et al. 2018).

From the environmental point of view, an appropriate destination of the generated wastes could represent a solution for reuse and add value to this product (Schwantes et al. 2018). In this sense, in order to adsorb contaminants from water, several authors have tried to develop biosorbents from agroindustrial residues (Schwantes et al. 2018; Olaoye et al. 2018; Alcaraz et al. 2018).

Among various pollutants that contaminate water, toxic metals are the ones that cause the greatest concern to the human population. It happens because the water can easily disseminate these components, inferring bioaccumulation and/or biomagnification in the food chain, damaging several organisms (Schiller et al. 2017).

PINUS CROP

Economic Aspects

Given the variations of the offer and demand of the timber sector, some forest companies choose to vary the production to enlarge their market niche. This strategy makes these companies more competitive becoming an alternative market for the farmers (David et al. 2017a).

In this sense, pinus is one of the main timber crops produced in several countries, such as Mexico (Martínez-Salvador et al. 2019), Brazil (IBA 2016), New Zealand, Australia, Chile, South Africa and Southwest Europe (EFGRP 2019).

Pinus reforestation emerges as an investment option that can help economic empowerment of small farmers in the generation of a new income. It is derived from another activity, thus reducing the exclusive financial dependence of traditional crops (Coelho and Coelho 2012).

The diversification of the energy matrix and new solutions for the rational and efficient use of natural resources are actions that can increase the supply of energy and meet the demand of countries such as Brazil (Oliveira et al. 2017).

In addition, in order to optimize the financial gain under the cultivation of the *Pinus taeda*, it is possible to state that the crop of 1600 plants ha^{-1} results in a better annual profit for the farmer, promoting maximization of profits (David et al. 2017b).

Social Aspects

Pinus crop provides valuable services for the population well-being (Shimizu et al. 2017). Its use is widespread in the pulp and paper industry, as well as in structures and finishes in civil construction, furniture manufacturing, packaging and energy biomass (Shimizu et al. 2017; Choudhury et al. 2014).

Due to this diversity of uses, this crop assists social strengthening, since it constitutes an income option allowing and aiding the maintenance of human in the countryside, avoiding the harmful effects of the rural exodus (Coelho and Coelho 2012).

In this way, the rural producer can be benefited from various forms of contractual pinus exploitation, especially: a) business pinus; b) partnership; c) fixed lease; d) lease for the exploration of pinus. In function of this, it is considered an option of development for the rural producer (Souza et al. 2004).

Environmental Aspects

The increase in the need for wood production for energy purposes has occurred to meet the generated demand by industry (wood blades, panels, mechanical processing, pulp and paper). These factors are preponderant to diversify the cultivation for timber purposes (Coelho and Coelho 2012).

In this aspect, it is emphasized that, among the species that contribute to this progress, there are those of the *Pinus* sp. genus (IBA 2016). In Brazil, for example, before the use of the pinus, the main source of raw material was Araucaria angustifolia (Coelho and Coelho 2012).

The pinus crop is very important to the environment. This is due to its characteristic of colonizing species. Seeds deposited on exposed soil have the ability to germinate in places without vegetation. Once established in the area, its roots develop and begin to absorb nutrients from deeper layers (Shimizu et al. 2017).

Planted forests offer numerous advantages to the environment. In these cases, it is not necessary to remove the soil (or inversion the soil) for the implantation of forests and the permanent cover provide significant edaphoclimatic benefits to the planet. In addition, it is emphasized the follow benefits (Souza et al. 2004): a) one hectare of planted forest offers in wood the equivalent of 30 hectares of native forest, thus reducing the pressure on deforestation. b) one hectare of eucalyptus absorbs about 10 tons of carbon from the atmosphere each year, contributing greatly to mitigating and slowing global warming and the greenhouse effect.

Reforestation with *Pinus* sp. genus can help to preserve native species due to the reduction of deforestation by anthropic actions (Choudhury et al. 2014).

CONTAMINATION OF WATER RESOURCES

The quality and quantity of water affects all ecosystems directly and indirectly. For this reason, the study of water resources has fundamental importance for the maintenance of life on planet (Samhan et al. 2017).

Although water is essential for all life on Earth, anthropic activities (agricultural and industrial) that generate effluents with high organic matter and metal are responsible to contaminate the soils and aquatic systems (Kuroda et al. 2018; Duncan et al. 2018). Once contaminated, water bodies have a negative effect on aquatic biota (Alrumman et al. 2016).

The unprecedented release of these elements into the environment is a dilemma for all living organisms. Metals in toxic levels have the ability to interact with several vital biomolecules, causing severe morphological, metabolic and physiological anomalies in plants, chlorosis of shoots to lipid peroxidation and protein degradation (Emamverdian et al. 2015).

Toxic Metals

The insertion of contaminants (organic and inorganic) into the water resources is responsible to deteriorate their quality. Among the pollution of inorganic sources, the contamination by metals is worrisome (Schiller et al. 2017). Metals are inserted in the environment by the weathering of the rocks, which is a very important process since it releases metals considered essential for fauna and flora (Blackmore et al. 2018).

As for the toxic effect for plants and animals, they are usually associated with the large amount of metals released into the environment by anthropic activities (Blackmore et al. 2018). The toxic effect occurs even in small quantities, causing bioaccumulation and biomagnification in the food chain (Schiller et al. 2017).

SUSTAINABLE ALTERNATIVES OF WATER RESOURCES REMEDIATION

Conventional Methods

Removal of contaminants from aqueous solutions has been a lot of studied in the field of research in recent years. In this context, several methods can be used in the remediation of water resources. Among them are conventional methods that use chemical, physical and biological techniques, which can be cited: coagulation and flocculation, oxidation or reduction, chemical precipitation, separation by membranes, among others that aim to reduce to a safe level the contamination of waters (Tchobanoglous et al. 2013).

However, these remediation techniques have disadvantages. Due to the high consumption of resources, they corroborate to the environmental emissions from the treatment process, as well as low efficiency.

Among the main techniques used to remove toxic metals from industrial effluents, chemical precipitation is the most used. However, this technique presents low efficiency in the treatment of very dilute solutions, especially when the concentration of metallic ions is between 1 to 100 mg L^{-1} (Wang and Chen 2009).

In this way, it is necessary a new approach for the remediation of contaminated water by metal (Taghizadeh et al. 2013). Therefore, the search for innovative and efficient methods in the removal of low concentrations of toxic metals at a lower cost becomes a challenge.

Innovative Methods

Innovative methods that enable the removal of low concentrations of contaminants at low cost are fundamental. The concept of sustainable remediation aims to eliminate contaminations using techniques that have low cost and without major impacts on the environment. Sustainable remediation avoids major stresses on species, habitats and ecosystems during remediation processes (Dunmade 2013; Bardos et al. 2011).

In this way, the sorption process is highlighted as a viable alternative. In this process, the occurrence of adsorption and absorption is observed simultaneously. This characteristic allows the removal of toxic metals concentrations and other contaminants in aqueous solutions (Gusmão et al. 2012) being extremely efficient and economically viable (Fu and Wang 2011).

The absorption process is characterized by the diffusion of the contaminant into the porous matrix of the solid. On the other hand, adsorption is the adhesion of ions or molecules to a solid surface. This process can be of a physical or chemical nature, depending on the type of interaction. In this context, the material accumulated at the interface is called adsorbate and the solid material is called adsorbent (Michalak et al. 2013; Gadd 2009).

Among the adsorbents that can be applied in this process are natural adsorbents, chemically modified adsorbents and activated carbons. It is emphasized that adsorption is one of the most used techniques to remove pollutants present in water systems (Foo 2010). The adsorption process has been widely researched for metal removal due to flexibility in design and operation, high removal capacity, cost-effectiveness and availability of different adsorbents. However, its efficiency is directly related to the type of adsorbent used (Pozdniakova et al. 2016).

Thus, in recent years the increase of researches that are dedicated to the study of new adsorbent materials has been noticed, with the objective of maximizing the removal capacity and selectivity in the removal of contaminants. In addition, the search for adsorbents from alternative materials, such as biosorbents, is also observed.

Natural Adsorbents

Natural adsorbents or biosorbents are materials from biological origin. Biosorption is a subcategory of adsorption (Michalak et al. 2013) and it can be defined as the removal/binding of desired substances from a solution by biological material. These substances may be organic or inorganic and soluble or insoluble (Gadd 2009).

A wide range of biomaterials has been used as biosorbents for the removal of contaminants. Materials of microbiological origin, originating from plants and animals, arouse great interest (Al-Masri 2010).

However, more recently some materials have stood out in the field of research. Among them, the use of agricultural residues, polysaccharides and industrial waste biomaterials (Witek-Krowiak and Reddy 2013; Reddy et al. 2012; Blázquez et al. 2011).

Due to the presence of a great variety of functional chelating groups, the biosorbents present high affinity for metallic ions (Volesky 2007). Among the natural materials that have been used and received attention may be mentioned: rice barks (Manique et al. 2012), coconut barks (Acheampong et al. 2013), bagasse of sugarcane (Khoramzadeh et al. 2013), seeds of *Moringa oleifera* (Meneghel et al. 2013), *Macadamia integrifólia* (Vilas Boas et al. 2016), wheat straw (Coelho et al. 2016), açaí endocarp (*Euterpe oleracea* M.) (Gonçalves Jr. et al. 2016), *Jatropha curcas* (Nacke et al. 2017; Nacke et al. 2016), macadamia (Honorato et al. 2017), pinus barks (Schwantes et al. 2018), cassava barks (Schwantes et al. 2017) among others.

It should be noted that these materials have the potential to be used in their natural form, chemically modified adsorbents or activated carbon.

Chemically Modified Adsorbents

The use of chemical modifications in biosorbents has been extensively studied in recent years. The use of suitable modifying agents is able to potentiate the adsorptive capacity of the biosorbent in the contaminants removal.

In this context, several studies have been conducted with the objective of evaluating the adsorptive potential of alternative materials submitted to chemical modifications. It is observed a large number of studies that are dedicated to the use of agroindustrial residues with different chemical treatments, among them: pinus bark (Schwantes et al. 2018), papaya seed (Yadav and Singh 2014), cassava bark (Schwantes et al. 2017) and bagasse of sugarcane (Santos et al. 2011).

This is justified by the potential that these chemical substances have to promote modifications in the adsorbent material. The use of chemical modification can cause changes such as introduction of functional groups in the adsorbent structure and changes in functional groups located directly on its surface (Demirbas 2008). It is also observed the increase in the number and dispersion of functional groups available for bonding metals.

In addition, the effect of the modifications on the biosorption properties may be the result of better contact surface characteristics of the biosorbent. It is verified the increase of the porosity and removal of initially adsorbed compounds (Rajczykowski et al. 2018). It is also able to improve chemical surface heterogeneity and increase and alter the adsorbent surface morphology (Anastopoulos et al. 2018).

It is important to emphasize that these changes are very important and they can corroborate with possible increase of the adsorption capacity of metals (Santos et al. 2010; Santos et al. 2011). It is valid to emphasize that when biomasses are chemically treated, the increase of the adsorption capacity usually occurs without raising the cost of the final product (Guyo et al. 2015).

REMEDIATION OF CONTAMINATED WATER USING PINUS BARK AS ADSORBENT IN THE REMEDIATION OF TOXIC METALS

Main Scientific Results

Obtainment and Characterization of Materials

The adsorbents were prepared using the external bark of *Pinus elliottii* trees obtained from an area of reforestation in the municipality of Chapecó – SC (south of Brazil). The barks were removed from three positions on the stem: base, middle and top, being mixed and homogenized.

After the collection, the material was sent to the Laboratory of Environmental and Instrumental Chemistry of the State University of the West of Paraná - UNIOESTE, Campus of Marechal Cândido Rondon - PR. Drying, homogenization and trituration of the pinus bark were carried out.

In the next step, modifying chemical agents were added in uniform particles, aiming the chemical activation of the material. Thus, solutions of hydrogen peroxide (H_2O_2), sulfuric acid (H_2SO_4) and sodium hydroxide (NaOH) were used in the concentration of 1 mol L^{-1}.

After the modification, four materials were originated, being them: pinus *in natura* (P. *in natura*), pinus chemically modified with H_2O_2 (P. H_2O_2), pinus chemically modified with H_2SO_4 (P. H_2SO_4) and pinus chemically modified with NaOH (P. NaOH).

In order to understand the characteristics of the adsorbents, the characterization were performed using the determination of chemical composition and techniques of scanning electron microscopy (SEM), infrared spectroscopy (FTIR) and point of zero charge (pH_{PZC}).

The concentration of the metals present in the pinus bark *in natura* and chemically modified materials with H_2O_2, H_2SO_4 and NaOH is shown in Table 1.

Table 1. Average contents of chemical elements
of the adsorbents P. *in natura* (Strey et al. 2013) and modified P. H_2O_2,
P. H_2SO_4 and P. NaOH (Schwantes et al. 2018)

Adsorbents	K	Ca	Mg	Cu	Zn	Mn	Fe	Cd	Pb	Cr
	g kg^{-1}			mg kg^{-1}						
P. *in natura*	3.05	2.69	0.67	4.50	35.00	155.00	1060.00	<LQ	22.00	4.10
P. H_2O_2	1.94	2.34	0.49	2.60	34.80	148.60	1054.31	<LQ	12.00	4.10
P. H_2SO_4	1.64	1.83	0.35	4.10	22.10	29.20	1001.58	<LQ	21.50	<LQ
P. NaOH	2.95	2.69	0.64	2.80	25.90	144.80	703.10	<LQ	10.20	0.20

LQ (limits of quantification): K = 0.01; Ca = 0.005; Mg = 0.005; Cu = 0.005; Fe = 0.01; Mn = 0.01; Zn = 0.005; Cd = 0.005; Pb = 0.01; Cr = 0.01.

Figure 1. SEM in 800 times of approximation of the adsorbents P. *in natura* (a) (Strey et al. 2013), and P. H_2O_2 (b), P. H_2SO_4 (c) and P. NaOH (d) (Schwantes et al. 2018).

The results showed a change in the concentration of K, Ca, Mg, Cu, Zn, Mn, Fe, Pb and Cr by H_2O_2, H_2SO_4 and NaOH solutions, which caused a decrease in the quantity of elements, modifying the structure and chemical composition of the adsorbent. This decrease in content occurs as the modifying solutions extract a certain amount of biomass from the adsorbents, besides the water rinses after the modification.

SEM (Figure 1) registered the structure present in the adsorbents. The surface characteristics observed in the *in natura* material are reproduced in the modified adsorbents, but with a greater number of cracks and pores.

The microstructures of the adsorbent P. *in natura* have a very irregular surface with large and significant voids, besides laminar structures with cracks and pores. Thus, according to Gonçalves Jr. et al. (2016), it is considered good adsorptive characteristics.

For the material P. H_2O_2 it can be observed a heterogeneity in the surface of the adsorbent in sponge format. Probably, this happens in function of the oxidant power of the modifying solution. The adsorbent P. H_2SO_4 showed irregular surface with many cracks as a function of the dehydrating action of its chemical agent.

The use of a strong base of high solubility has caused irregular and heterogeneous surface with cavities in the material P. NaOH. These characteristics are obtained with modification by simple acids and bases. These agent products help in the formation of a residual product of low cost in the process of toxic metals adsorption in aqueous solutions, due to the heterogeneity of the surface of these adsorbents (Schwantes et al. 2016).

Infrared spectra in the range of 500-4000 cm^{-1} demonstrate the presence of several functional groups on the surface of the adsorbents (Figure 2), which may favor the adsorption process. By several wave numbers, it was found similar vibrational stretches in all the materials coming from pinus. This fact evidences that there are several functional groups that are present in *P. in natura* and that even with the chemical modifications, they remained present, such as primary and secondary amine, amides, hydroxyl groups present in cellulose and lignin, alkanes, aromatic structures, phenols, among others.

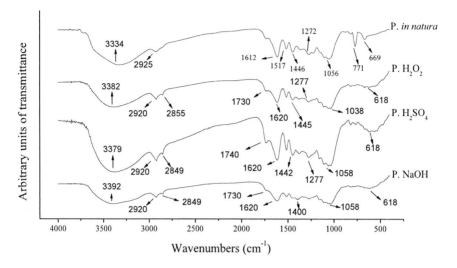

Figure 2. Infrared Spectra for adsorbents based on pinus bark *in natura* (Strey et al. 2013) and modified with H_2O_2, H_2SO_4 and NaOH (Schwantes et al. 2018).

Therefore, the modified materials presented new bands in relation to the precursor material. According to Stuart (2004), these bands may refer to stretches of C-H and C=O, related to aldehydes and carboxylic groups, respectively.

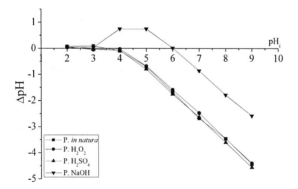

Figure 3. Point of zero charge (pH$_{PZC}$) of the adsorbents P. *in natura* (Strey et al. 2013) and modified P. H_2O_2, P. H_2SO_4 and P. NaOH (Schwantes et al. 2018).

The results obtained for the point of zero charge (pH$_{PZC}$), after treatment with the modifying solutions, indicated that the values for P. *in natura*, H_2O_2, P. H_2SO_4 and P. NaOH were altered, being respectively

3.50, 2.50, 2.47 and 6.03 (Figure 3). This change was already expected, since the variation of pH_{PZC} occurred according to the alkalization or acidification power of each solution. Thus, the material P. NaOH presents preference in anion adsorption, while the other adsorbents prefer cations, such as Cd^{2+}, Pb^{2+} e Cr^{3+}. This preference was observed since the results of the pH_{PZC} were lower than the pH of the solution (Mimura et al. 2010).

After adsorption, adsorbents were subjected to an influence multivariable analysis of the adsorbent dose and the pH of the solution fortified with the Cd^{2+}, Pb^{2+} and Cr^{3+} metals. For that, the Central Composite Rotational Design (CCRD) was used, which allows to determine the influence of each one of the variables and the possible interaction between them. This analysis generates an empirical and quadratic mathematical model with validity in the experimentally tested range (Barros et al. 2010). The doses of the adsorbents in the range of 5 to 25 g L^{-1} were evaluated while the pH was evaluated in the range 3.00 to 7.00.

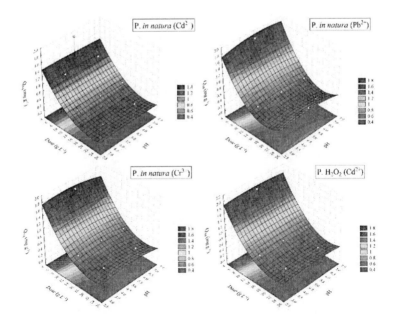

Figure 4. (Continued).

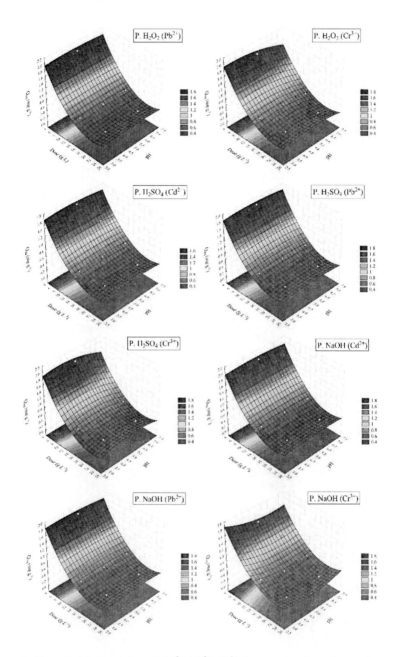

Figure 4. Capacity of adsorption of Cd^{2+}, Pb^{2+}, Cr^{3+} from contaminated water by the materials.

According to the results, it was possible to develop the surface response graphs (Figure 4). In these graphs it can be observed that the highest adsorbed quantities of the three contaminants (Cd^{2+}, Pb^{2+} and Cr^{3+}) occurred when there was less availability of adsorbent material in the solution contaminated, i.e., 4 g L^{-1} or 200 mg. However, the pH of the solution did not influence the amount of toxic metals adsorbed (Strey et al. 2013; Schwantes et al. 2018).

The studied materials were submitted to linear mathematical models of kinetics, equilibrium and thermodynamics, in order to verify the best material to be used in the removal of these contaminants. In the next topics will be presented the best adsorbents for each toxic metal studied.

Removal of Cadmium (Cd^{2+})

According to the mathematical models used (Figure 5), the best material for the removal of Cd^{2+} was P. NaOH, with an increase in the adsorption rate of 58, 68 and 33% when compared to the materials P. *in natura*, P. H_2O_2 and P. H_2SO_4, respectively (Strey et al. 2013; Schwantes et al. 2018).

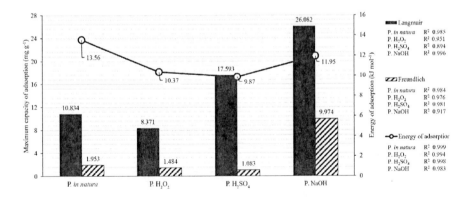

Figure 5. Adsorption capacity and energy of materials for removal of metals Cd^{2+} from contaminated water.

The Langmuir model presented the following removal values: 10.83 mg g^{-1} for P. *in natura* (Strey et al. 2013), 8.37 mg g^{-1} for P. H_2O_2, 17.59 mg g^{-1} for P. H_2SO_4 and 26.08 mg g^{-1} for P. NaOH (Schwantes et al. 2018).

The values of R^2 were close to one, indicating satisfactory adjustments for both models (Langmuir and Freundlich). In this way, the adsorption occurs in mono and multilayer.

The adsorption process indicates that a high chemical attraction occurred between adsorbent and adsorbate particles, presenting sorption energy values greater than 8 kJ mol^{-1} (Figure 5). This data corroborates to the results obtained by the studied kinetic models (Table 2), where there was only adjustment by the pseudo-second order model, with values of Q_{eq} (exp.) and Q_{eq} (calc.) close to each other, suggesting occurrence of chemisorption (chemical adsorption) (Strey et al. 2013; Schwantes, 2016).

Table 2. Kinetic parameters for Cd^{2+} removal by adsorbents based on pinus bark

Parameters/Adsorbents	P. *in natura*	P. H_2O_2	P. H_2SO_4	P. NaOH
Pseudo-first order				
K_1 (min^{-1})	-0.011	-0.011	-0.017	-0.004
$Q_{eq\ (calc.)}$ (mg g^{-1})	0.105	0.057	0.042	0.258
R^2	0.715	0.744	0.733	0.337
Pseudo-second order				
K_2 (g mg^{-1} min^{-1})	0.294	2.672	-1.308	-1.082
$Q_{eq\ (calc.)}$ (mg g^{-1})	1.188	2.270	2.025	2.433
R^2	0.999	0.999	0.992	0.999
$Q_{eq\ (exp.)}$ (mg g^{-1})	1.182	2.261	2.036	2.444

K_1: constant rate of pseudo-first order; Q_{eq}: amounts of adsorbate retained per gram of adsorbent at equilibrium; R^2: adjusted coefficient of determination; K_2: constant rate of pseudo-second order.

Removal of Lead (Pb^{2+})

Among the results presented by the mathematical models for Pb^{2+} removal, the material P. NaOH obtained the highest adsorption rate among the adsorbents (44.66 mg g^{-1}). As can be seen in Figure 6, Pb^{2+} removal rates suggest the predominance of monolayer adsorption according to the excellent adjustments of R^2 for Langmuir model (Schwantes et al. 2018).

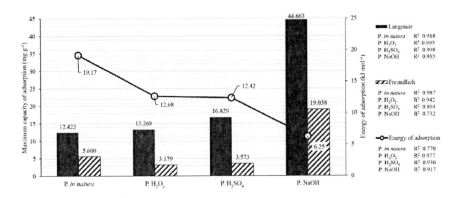

Figure 6. Adsorption capacity and energy of materials for removal of metals Pb^{2+} from contaminated water.

Modification of the material with NaOH caused the removal of Pb^{2+} from water an increase of 359.6% when compared to P. *in natura*, which removed 12.42 mg of Pb^{2+} per g of adsorbent (Schwantes et al. 2018).

When comparing the values obtained in this work with other studies such as Schwantes et al. (2015), Schwantes et al. (2016) and Coelho et al. (2018), it is possible to observe that chemical modification of the material potentiates the adsorption capacity.

Table 3. Kinetic parameters for Pb^{2+} removal by adsorbents based on pinus bark

Parameters/Adsorbents	P. *in natura*	P. H_2O_2	P. H_2SO_4	P. NaOH
Pseudo-first order				
K_1 (min^{-1})	-0.006	-0.016	-0.006	-0.034
$Q_{eq\,(calc.)}$ (mg g^{-1})	0.016	0.094	0.166	0.182
R^2	0.121	0.845	0.640	0.726
Pseudo-second ordem				
K_2 (g mg^{-1} min^{-1})	1.337	0.357	0.281	-0.449
$Q_{eq\,(calc.)}$ (mg g^{-1})	1.248	2.550	2.500	2.359
R^2	1.000	0.999	0.998	0.996
$Q_{eq\,(exp.)}$ (mg g^{-1})	1.250	2.499	2.428	2.375

K_1: constant rate of pseudo-first order; Q_{eq}: amounts of adsorbate retained per gram of adsorbent at equilibrium; R^2: adjusted coefficient of determination; K_2: constant rate of pseudo-second order.

The values of R^2 (Table 3) show that the model that had better explain the adsorption kinetics of Pb^{2+} is pseudo-second order model. The other models are not satisfactory. To corroborate these results, the $Q_{eq\ (calc.)}$ for the pseudo-second order, for all cases, is very close to $Q_{eq\ (exp.)}$, unlike the pseudo-first order. This is an indication that the process-limiting step is chemisorption (with chemical bonds involving valence and electron sharing forces between adsorbent and adsorbate) (Feng et al. 2011).

Removal of Chromium (Cr³⁺)

The modified adsorbent P. H_2SO_4 presented adsorption of Cr^{3+} mainly in multilayers, whereas the materials P. H_2O_2 and P. NaOH presented Cr^{3+} adsorption mainly in monolayers. Material P. *in natura* showed both monolayer and multilayer adsorption, according to Langmuir and Freundlich parameters (Figure 7) (Strey, 2013; Schwantes, 2016; Schwantes et al. 2018).

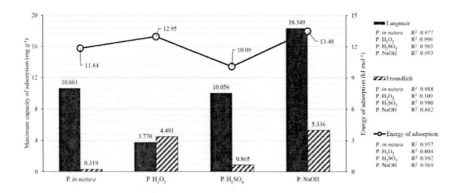

Figure 7. Adsorption capacity and energy of materials for removal of metals Cr^{3+} from contaminated water.

The highest adsorbed amount of Cr^{3+} was found using the material P. NaOH, (18.34 mg g^{-1}), where suggestively the chemisorption process occurred, with the best result of R^2 for pseudo-second order (Table 4). The adsorbents P. H_2O_2 and P. H_2SO_4 obtained satisfactory adjustments for pseudo-first order with values of R^2 close to one, suggesting that the adsorption phenomenon can be a physical process. This fact only occurred for Cr^{3+} adsorption (Strey, 2013; Schwantes et al. 2018).

Table 4. Kinetic parameters for Cr^{3+} removal by adsorbents based on pinus bark

Parameters/Adsorbents	P. *in natura*	P. H_2O_2	P. H_2SO_4	P. NaOH
Pseudo-first order				
K_1 (min^{-1})	0.001	-0.009	-0.028	0.004
$Q_{eq (calc.)}$ (mg g^{-1})	1.229	0.356	0.569	0.078
R^2	0.930	0.870	0.901	0.233
Pseudo-second order				
K_2 (g mg^{-1} min^{-1})	0.086	0.126	0.161	-1.196
$Q_{eq (calc.)}$ (mg g^{-1})	1.191	1.906	1.862	2.030
R^2	0.999	0.998	0.998	0.998
$Q_{eq (exp.)}$ (mg g^{-1})	1.142	1.745	1.718	2.061

K_1: constant rate of pseudo-first order; Q_{eq}: amounts of adsorbate retained per gram of adsorbent at equilibrium; R^2: adjusted coefficient of determination; K_2: constant rate of pseudo-second order.

CONCLUSION

It possible to conclude that pinus is a crop of economic, social and environmental importance for several countries. In addition, the use of this agroindustrial co-product (*Pinus elliottii* bark) as an adsorbent material constitutes a sustainable practice. This practice complements the final productive stages of this species, transforming a residue (with potential environmental impact) into adsorbent material with potential for remediation of contaminated water.

The results found in literature shown that the use of pinus bark is an excellent alternative for toxic metal removal. However, it is important to emphasize that the adsorption process is influenced by chemical modifications. Since, the modified materials shown better results when compared to the *in natura* material.

It is worth mentioning that the use of the pinus bark as an adsorbent has the advantage of being easily found, accessible, and abundant and it presents low cost. These characteristics make its applications a viable practice for the remediation of contaminated water contaminated with Cd^{2+}, Pb^{2+} and Cr^{3+}.

REFERENCES

Acheampong, M. A., Pakshirajan, K., Annachhatre, A. P., and Lens, P. N. L. 2013. "Removal of Cu (II) by biosorption onto coconut shell in fixed-bed column systems." *Journal of Industrial and Engineering Chemistry* 19:841-48. doi:10.1016/j.jiec.2012.10.029.

Alcaraz, L., Fernández, A. L., García-Díaz, I., and López, F. A. 2018. "Preparation and characterization of activated carbons from winemaking wastes and their adsorption of methylene blue." *Adsorption Science & Technology* 36:1331-51. doi:10.1177/0263617418770295.

Al-Masri, M. S., Amin, Y., Al-Akel, B., and Al-Naama, T. 2010. "Biosorption of cadmium, lead, and uranium by powder of poplar leaves and branches." *Applied biochemistry and biotechnology* 160:976-87. doi:10.1007/s12010-009-8568-1.

Alrumman, S. A., El-kott, A. F., and Keshk, S. M. A. S. 2016. "Water pollution: source & treatment." *American Journal Environmental Engeneering* 6:88–98. doi:10.5923/j.ajee.20160603.02.

Anastopoulos, I., Robalds, A., Tran, H. N., Mitrogiannis, D., Giannakoudakis, D. A., Hosseini-Bandegharaei, A., and Dotto, G. L. 2018. "Leaf Biosorbents for the Removal of Heavy Metals." In *Green Adsorbents for Pollutant Removal*, edited by G. Crini, and E. Lichtfouse, 87-126. Cham: Springer. doi:10.1007/978-3-319-92162-4_3.

Bardos, P., Bone, B., Boyle, R., Ellis, D., Evans, F., Harries, N. D., and Smith, J. W. 2011. "Applying sustainable development principles to contaminated land management using the SuRF-UK framework." *Remediation Journal* 21:77-100. doi:10.1002/rem.20283.

Barros, N. B., Bruns, R. E., and Scarminio, I. S. 2010. *How do experiments-applications in science and industry*. Porto Alegre: Bookman.

Blackmore, S., Vriens, B., Sorensen, M., Power, I. M., Smith, L., Hallam, S. J., Mayer, K. U., and Beckie, R. D. 2018. "Microbial and geochemical controls on waste rock weathering and drainage quality." *Science of the Total Environment* 640:1004–14. doi:10.1016/j. scitotenv.2018.05.374.

Blázquez, G., Martín-Lara, M. A., Tenorio, G., and Calero, M. 2011. "Batch biosorption of lead (II) from aqueous solutions by olive tree pruning waste: Equilibrium, kinetics and thermodynamic study." *Chemical Engineering Journal* 168:170-77. doi:10.1016/j.cej. 2010.12.059.

Choudhury, H., Kumaria. S., and Tandon, P. 2014. "Pinus Biotechnology: Progress and Prospects." In *Tree Biotechnology*, edited by K. G. Ramawat, J. M. Merillon, and M. R. Ahuja, 1-27. Boca Raton: CRC Press. doi:10.1201/b16714-10.

Coelho, G. F., Gonçalves Jr, A. C., Nóvoa-Muñoz, J. C., Fernández-Calviño, D., Arias-Estévez, M., Fernández-Sanjurjo, M. J., Álvarez-Rodríguez, E., and Núñez-Delgado, A. 2016. "Competitive and non-competitive cadmium, copper and lead sorption/desorption on wheat straw affecting sustainability in vineyards." *Journal of cleaner production* 139:1496-1503. doi:10.1016/j.jclepro.2016.09.021.

Coelho, G. F., Gonçalves, A. C., Schwantes, D., Rodríguez, E. Á., Tarley, C. R. T., Dragunski, D., and Junior, É. C. 2018. "Removal of Cd (II), Pb (II) and Cr (III) from water using modified residues of *Anacardium occidentale* L." *Applied Water Science* 8:96. doi:10.1007/s13201-018-0724-8.

Coelho, M. H., and Coelho, M. R. F. 2012. "Potencialidades econômicas de florestas plantadas de *Pinus elliottii*, em pequenas propriedades rurais." *Revista Paranaense de Desenvolvimento* 123:257-278. ["Economic Potential of Planted Forests of *Pinus elliottii* in Small Rural Properties." *Paranaense Journal of Development* 123:257-278].

David, H. C., Arce, J. E., Oliveira, E. B., Netto, S. P., Miranda, R. O. V., and Ebling, Â. A. 2017b. "Economic analysis and revenue optimization in management regimes of *Pinus taeda*." *Revista Ceres* 64:222–231. doi:10.1590/0034-737x201764030002.

David, H. C., Péllico Netto, S., Arce, J. E., Corte, A. P. D., Marinheski Filho, A., and Araújo, E. J. G. 2017a. "Efeito da Qualidade do Sítio e do Desbaste na Produção de Pinus." *Floresta e Ambiente* 24. doi:10.1590/2179-8087.096414. ["Effect of Site Quality and Thinning in Pine Yield." *Forest and Environment* 24].

Delucius, R., Santos, P. S. B., Beltrame, R., and Gatto, D. A. 2018. "Propriedades químicas e energéticas de resíduos florestais provenientes de plantações de pinus." *Revista Árvore* 41. doi:10.1590/1806-90882017000500007. ["Chemical and fuel properties of forestry wastes from pine plantations." *Tree Journal* 41].

Demirbas, A. 2008. Heavy metal adsorption onto agro-based waste materials: a review. *Journal of Hazardous Materials* 157:220-29. doi:10.1016/j.jhazmat.2008.01.024.

Duncan, A. E., De Vries, N., and Nyarko, K. B. 2018. "Assessment of Heavy Metal Pollution in the Sediments of the River Pra and Its Tributaries." *Water, Air, & Soil Pollut*ion 229:1-10. doi:10.1007/s11270-018-3899-6.

Dunmade, I. 2013. "The Role of Sustainable Remediation in the Preservation of Biodiversity: Areas of Opportunities." *Journal of Biodiversity Management & Forestry* 2. doi:10.4172/2327-4417.1000104.

EFGRP - European Forest Genetic Resousces Programme. 2019. "Genetic diversity is the basis of resilience." Accessed March 21. http://www.euforgen.org/species/Pinus-radiata/.

Emamverdian, A., Ding, Y., Mokhberdoran, F., & Xie, Y. (2015). "Heavy Metal Stress and Some Mechanisms of Plant Defense Response." *Scientific World Journal* 2015:1–18. doi:10.1155/2015/756120.

Feng, N., Guo, X., Liang, S., Zhu, Y., and Liu, J. 2011. "Biosorption of heavy metals from aqueous solutions by chemically modified orange peel." *Journal of Hazardous Materials* 185:49-54. doi:10.1016/j. jhazmat.2010.08.114.

Foo, K. Y., and Hameed, and Bassim H. 2010. "An overview of dye removal via activated carbon adsorption process." *Desalination and Water Treatment* 19:255-274. doi:10.5004/dwt.2010.1214.

Fu, F., and Wang, Q. 2011. "Removal of heavy metal ions from wastewaters: a review." *Journal of Environmental Management* 92:407-18. doi:10.1016/j.jenvman.2010.11.011.

Gadd, G. M. 2009. "Biosorption: critical review of scientific rationale, environmental importance and significance for pollution treatment." *Journal of Chemical Technology & Biotechnology: International Research in Process, Environmental & Clean Technology* 84:13-28. doi:10.1002/jctb.1999.

Gonçalves Jr, A. C., Coelho, G. F., Schwantes, D., Rech, A. L., Campagnolo, M. Â., and Miola Jr, A. 2016. "Biosorption of Cu (II) and Zn (II) with acai endocarp *Euterpe oleracea* M. in contaminated aqueous solution." *Acta Scientiarum. Technology* 38:361-71. doi:10. 4025/actascitechnol.v38i3.28294.

Gusmão, K. A. G., Gurgel, L. V. A., Melo, T. M. S., and Gil, L. F. 2012. "Application of succinylated sugarcane bagasse as adsorbent to remove methylene blue and gentian violet from aqueous solutions– kinetic and equilibrium studies." *Dyes and Pigments* 92:967-74. doi:10.1016/j.dyepig.2011.09.005.

Guyo, U., Mhonyera, J., and Moyo, M. 2015. "Pb (II) adsorption from aqueous solutions by raw and treated biomass of maize stover–a comparative study." *Process Safety and Environmental Protection* 93:192-200. doi:10.1016/j.psep.2014.06.009.

Honorato, A. C., Pardinho, R. B., Dragunski, D. C., Junior, A. C. G., and Caetano, J. 2017. "Biosorbent of macadamia residue for cationic dye adsorption in aqueous solution." *Acta Scientiarum. Technology* 39:97-102. doi:10.4025/actascitechnol.v39i1.28524.

IBA – Indústria Brasileira de Árvores. 2016. "*Anuário estatístico da IBA.*" Accessed March 15. http://iba.org/images/shared/Biblioteca/IBA_

RelatorioAnual2016_.pdf. [IBA – Brazilian Tree Industry. 2016. "*IBA Statistical Yearbook*"].

Khoramzadeh, E., Nasernejad, B., and Halladj, R. 2013. "Mercury biosorption from aqueous solutions by sugarcane bagasse." *Journal of the Taiwan Institute of Chemical Engineers* 44:266-269. doi:10.1016/ j.jtice.2012.09.004.

Kuroda, M., Hara, K., Takekawa, M., Uwasu, M., and Ike, M. 2018. "Historical Trends of Academic Research on the Water Environment in Japan: Evidence from the Academic Literature in the Past 50 Years." *Water* 10:1-15. doi:10.3390/w10101456.

Manique, M. C., Faccini, C. S., Onorevoli, B., Benvenutti, E. V., and Caramão, E. B. 2012. "Rice husk ash as an adsorbent for purifying biodiesel from waste frying oil." *Fuel* 92:56-61. doi:10.1016/j.fuel. 2011.07.024.

Martínez-Salvador, M., Mata-Gonzalez, R., Pinedo-Alvarez, S., Morales-Nieto, C. R., Prieto-Amparán, J. A., Vázquez-Quintero, G., Villarreal-Guerrero., and F. 2019. "A Spatial Forestry Productivity Potential Model for *Pinus arizonica* Engelm, a Key Timber Species from Northwest Mexico." *Sustainability* 11:829. doi:10.3390/su11030829.

Meneghel, A. P., Gonçalves, A. C., Rubio, F., Dragunski, D. C., Lindino, C. A., and Strey, L. 2013. "Biosorption of cadmium from water using Moringa (*Moringa oleifera* Lam.) seeds." *Water, Air, & Soil Pollution* 224:1383. doi:10.1007/s11270-012-1383-2.

Michalak, I., Chojnacka, K., and Witek-Krowiak, A. 2013. "State of the art for the biosorption process—a review." *Applied Biochemistry and Biotechnology* 170:1389-1416. doi:10.1007/s12010-013-0269-0.

Mimura, A. M. S., Vieira, T. V. D. A., Martelli, P. B., and Gorgulho, H. D. F. 2010. "Utilization of rice husk to remove Cu^{2+}, Al^{3+}, Ni^{2+} and Zn^{2+} from wastewater." *Química Nova* 33:1279-1284. doi:10.1590/S0100-40422010000600012.

Nacke, H., Gonçalves, A. C., Campagnolo, M. A., Coelho, G. F., Schwantes, D., dos Santos, M. G., Briesch Jr, D. L., and Zimmermann, J. 2016. "Adsorption of Cu (II) and Zn (II) from Water by *Jatropha curcas* L. as Biosorbent." *Open Chemistry* 14:103-117. doi:10.1515/chem-2016-0010.

Nacke, H., Gonçalves, A. C., Coelho, G. F., Schwantes, D., Campagnolo, M. A., Leismann, E. A. V., Conradi Junior, E., and Miola, A. J. 2017. "Removal of Cd (II) from water using the waste of jatropha fruit (*Jatropha curcas* L.)." *Applied Water Science* 7:3207-22. doi:10.1007/s13201-016-0468-2.

Olaoye, R. A., Afolayan, O. D., Mustapha, O. I., and Adeleke, O. G. H. 2018. "The Efficacy of Banana Peel Activated Carbon in the Removal of Cyanide and Selected Metals from Cassava Processing Wastewater." *Advances in Research.* 16:1-12. doi:10.9734/AIR/2018/43070.

Oliveira, L. H., Barbosa, P. V. G., Lima, P. A. F., Yamaji, F. M., Sette Júnior, C. R. 2017. "Aproveitamento de resíduos madeireiros de *Pinus* sp. com diferentes granulometrias para a produção de briquetes." *Revista de Ciências Agrárias* 40:683-91. doi:10.19084/RCA17010. ["Use of wood residues of *Pinus* sp. with different granulometry to briquettes production." *Journal of Agrarian Sciences* 40:683-91].

Pozdniakova, T. A., Mazur, L. P., Boaventura, R. A., and Vilar, V. J. 2016. "Brown macro-algae as natural cation exchangers for the treatment of zinc containing wastewaters generated in the galvanizing process." *Journal of cleaner production* 119:38-49. doi:10.1016/j.jclepro.2016.02.003.

Rajczykowski, K., Sałasińska, O., and Loska, K. 2018. "Zinc Removal from the Aqueous Solutions by the Chemically Modified Biosorbents." *Water, Air, & Soil Pollution* 229:6. doi:/10.1007/s11270-017-3661-5.

Reddy, D. H. K., Lee, S. M., and Seshaiah, K. 2012. "Biosorption of toxic heavy metal ions from water environment using honeycomb biomass—an industrial waste material." *Water, Air, & Soil Pollution* 223:5967-82. doi:10.1007/s11270-012-1332-0.

Samhan, F. A., Elliethy, M. A., Hemdan, B. A., Youssef, M., and El-Taweel, G. E. 2017. "Bioremediation of oil-contaminated water by bacterial consortium immobilized on environment-friendly biocarriers." *Journal of Egyptian Public Health Association* 92:44-51. doi:10.21608/epx.2017.7009.

Santos, V. C. G., Tarley, C. R. T., Caetano, J., and Dragunski, D. C. 2010. "Assessment of chemically modified sugarcane bagasse for lead adsorption from aqueous medium." *Water Science and Technology* 62:457-65. doi:10.2166/wst.2010.291.

Santos, V. C., Souza, J. V., Tarley, C. R., Caetano, J., and Dragunski, D. C. 2011. "Copper ions adsorption from aqueous medium using the biosorbent sugarcane bagasse *in natura* and chemically modified." *Water, Air, & Soil Pollution* 216:351-59. doi:10.1007/s11270-010-0537-3.

Schiller, A. P., Schwantes, D., Gonçalves Jr, A. C., Manfrin, J., Klais, B., Parrales, A., and Kunh, A. 2017. "Teores de metais em cursos hídricos de Toledo - PR." *Revista de Ciências Ambientais* 11:53-70. doi:10.18316/rca.v11i3.3139. ["Metal content in water resources of Toledo – PR." *Journal of Environmental Sciences* 11:53-70].

Schwantes, D. 2016. *Preparo de biomassas vegetais modificadas quimicamente e aplicação em estudos adsortivos de Cd(II), Pb(II) e Cr(III).* PhD thesis. Universidade Estadual do Oeste do Paraná. [*Preparation of chemically modified plant biomasses and application in adsorption studies of Cd(II), Pb(II) and Cr(III).* PhD thesis. State University of Western Paraná].

Schwantes, D., Gonçalves Jr, A. C., Campagnolo, M. A., Tarley, C. R. T., Dragunski, D. C., de Varennes, A., Silva, A. K. S., and Conradi Junior, E. 2018. "Chemical modifications on *Pinus* bark for adsorption of toxic metals." *Journal of Environmental Chemical Engineering* 6:1271-78. doi:10.1016/j.jece.2018.01.044.

Schwantes, D., Gonçalves Jr, A. C., Campagnolo, M. A., Tarley, C. R. T., Dragunski, D. C., Manfrin, J., and Schiller, A. D. P. 2017. "Use of Co-Products from the Processing of Cassava for the Development of Adsorbent Materials Aiming Metal Removal." In *Cassava*, edited by V. Waisundara. London: IntechOpen. doi:10.5772/intechopen.71048.

Schwantes, D., Gonçalves, A. C., Coelho, G. F., Campagnolo, M. A., Dragunski, D. C., Tarley, C. R. T., Miola, A. J., and Leismann, E. A.

V. 2016. "Chemical modifications of cassava peel as adsorbent material for metals ions from wastewater." *Journal of Chemistry* 2016:1-15. doi:10.1155/2016/3694174.

Schwantes, D., Gonçalves, J., Coelho, G. F., Campgnolo, M. A., Santos, M. G., Miola Jr, A., and Leismann, E. A. V. 2015. "Crambe pie modified for removal of cadmium, lead and chromium from aqueous solution." *International Journal of Current Research* 7:21658-21669.

Shimizu, J. Y., Aguiar, A. V., Oliveira, E. B., Mendes, C. J., Murara Junior, M. I., Sousa, V. A., and Degenhardt-Goldbach, J. 2017. *Projeto Cooperativo de Melhoramento de Pínus - PCMP*. Colombo: Embrapa Florestas. [*Cooperative Pine Improvement Project – PCMP*. Colombro: Embrapa Forest].

Souza, A., Kreuz, C., and Motta, C. 2004. "Análise de empreendimentos florestais (pinus) como alternativa de renda para o produtor rural na região dos Campos de Palmas." *Revista de Administração da UFLA* 6:8-21. ["Analysis on forest entrepreneurship (pinus) as an alternative income source for the rural producer in the Campos de Palmas region." *UFLA Journal of Management* 6:8-21].

Strey, L. 2013. *"Biossorção de íons metálicos em águas utilizando casca de Pinus como material adsorvente alternativo."* Msc dissertation, Universidade Estadual do Oeste do Paraná. [*"Biosorption of metal ions in water using bark of pinus as an alternative adsorbent material"*. Msc dissertation, State University of Western Paraná].

Strey, L., Gonçalves Jr, A. C., Schwantes, D., Coelho, G. F., Nacke, H., and Dragunski, D. C. 2013. "Kinetics, equilibrium and thermodynamics of cadmium adsorption by a biosorbent from the bark of *Pinus elliottii*." In *Green Design, Materials and Manufacturing Processes*, edited by H. Bartolo, P. J. S. Bartolo, N. M. F. Alves, A. J. Mateus, H. A. Almeida, A. C. S. Lemos, F. Craveiro, C. Ramos, I. Reis, L. Durão, T. Ferreira, J. P. Duarte, F. Roseta, E. C. Costa, F. Quaresma, and J. P. Neves. Lisboa: CRC Press.

Stuart, B. H. 2004. *Infrared Spectroscopy: Fundamentals and Applications*. New York: John Wiley and Sons.

Taghizadeh, M., Kebria, D. Y., Darvishi, G., and Kootenaei, F. G. 2013. "The use of nano zero valent iron in remediation of contaminated soil and groundwater." *International Journal of Scientific Research in Environmental Sciences* 1:152-57. doi:10.12983/ijsres-2013-p152-157.

Tchobanoglous, G., Stensel, H. D., Tsuchihashi, R., Burton, F., Abu-Orf, M., Bowden, G., and Pfrang, W. 2013. *Wastewater engineering: treatment and resource recovery*. New York: McGraw-Hill Education.

Vilas Boas, N., Casarin, J., Gerola, G. P., Tarley, C. R. T., Caetano, J., Gonçalves Jr, A. C., and Dragunski, D. C. 2016. "Evaluation of kinetic and thermodynamic parameters in adsorption of lead (Pb^{2+}) and chromium (Cr^{3+}) by chemically modified macadamia (*Macadamia integrifolia*)." *Desalination and Water Treatment* 57:17738-17747. doi:10.1080/19443994.2015.1085906.

Volesky, B. (2007). "Biosorption and me." *Water Research* 41:4017-4029. doi:10.1016/j.watres.2007.05.062.

Wang, J., and Chen, C. 2009. "Biosorbents for heavy metals removal and their future." *Biotechnology Advances* 27:195-226. doi:10.1016/j.biotechadv.2008.11.002.

Witek-Krowiak, A., and Reddy, D. H. K. 2013. "Removal of microelemental Cr (III) and Cu (II) by using soybean meal waste–unusual isotherms and insights of binding mechanism." *Bioresource Technology* 127:350-357. doi:10.1016/j.biortech.2012.09.072.

Yadav, S. K., Singh, D. K., and Sinha, S. 2014. "Chemical carbonization of papaya seed originated charcoals for sorption of Pb (II) from aqueous solution." *Journal of Environmental Chemical Engineering* 2:9-19. doi:10.1016/j.jece.2013.10.019.

INDEX

Related Nova Publications

PHYTOCHEMICALS: PLANT SOURCES AND POTENTIAL HEALTH BENEFITS

EDITOR: Iman Ryan

SERIES: Plant Science Research and Practices

BOOK DESCRIPTION: The opening chapter of *Phytochemicals: Plant Sources and Potential Health Benefits* discusses macronutrients and micronutrients from plants along with their benefits to human health.

HARDCOVER ISBN: 978-1-53615-478-8
RETAIL PRICE: $230

PLANT DORMANCY: MECHANISMS, CAUSES AND EFFECTS

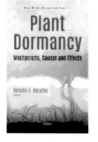

EDITOR: Renato V. Botelho

SERIES: Plant Science Research and Practices

BOOK DESCRIPTION: Dormancy is a mechanism found in several plant species developed through evolution, which allows plants to survive in adverse conditions and ensure their perpetuation. This mechanism, however, can represent a barrier that can compromise the development of the species of interest, and therefore, the success of its cultivation.

HARDCOVER ISBN: 978-1-53615-380-4
RETAIL PRICE: $160

To see a complete list of Nova publications, please visit our website at www.novapublishers.com